IN EXHIBITION AND DISPLAY

展览导视 II

（日）前田丰 编　常文心 译

辽宁科学技术出版社

PREFACE »

前言

Yutaka Maeda

Yutaka Maeda, art director and founder of ujidesign, Tokyo Japan. He has been working across various media, such as paper work, interface of website, exhibition, grand space of commercial complex. He creates visual identity for each project and achieves a successful design solution for the clients. Maeda has won many prizes with his rich idea and creativity, including 2014 Diamond award of SDA awards (Japan Sign Design Association), 2012 Grand prix of SDA.

前田丰

ujidesign 设计公司（日本东京）艺术总监、创始人。他的设计遍布各种媒介，包含纸面、网站界面、展览、大型商业空间等。他为每个项目打造合适的视觉识别系统，取得了令客户满意的成果。前田丰凭借丰富的理念和创新获得了大量设计奖项，其中包括 2014 日本标识设计协会钻石奖和 2012 日本标识设计协会最高奖。

I find the category of environmental graphic design fascinating. To begin with, the scale of the work is just so large. I do a lot of graphic-based design work, but environmental graphic design is by far the largest in terms of actual scale. The large size means that when you are choosing elements for the layout then you naturally have greater choice. It is, after all, possible to fill an extremely large space with one photo or 1,000.

In contrast, if you are working for a book or screen, then size of that medium itself limits the size of any object you can use. You don't have to really think about this with environmental graphic design. You can use the space available freely. For this reason, if you are making an emotional design then it can be made even more emotional, or if you're making a highly detailed design then it can be made even more so. The design can be more of whatever it is.

But of course, the wide range of options can itself pose problems. With small-scale design, like that of a book or leaflet, you can experience the design onscreen at more or less actual size. With environmental graphic design that is not possible. All you can do with environmental graphic design is print out one part of the object at full-size and then extrapolate what it would look like in your mind. Of course, you can also use a projector to project the design on a wall at full size. The point is that more testing is required.

Unlike designing for paper, there is also a greater range of materials to choose from. And each time you use a new material, you also require testing of new theories, layouts, colours, font sizes and so on.

The new graphics you produce must be faultless — with nothing disconcerting about them — and that's why they require such testing. Then again, the more time you've put into the work, then the happier you are when you finish.

Meanwhile, thorough testing also has the added benefit of resulting in more new discoveries. For that reason, I think you could say that environmental graphic design tends to encourage new work that pushes you out of your comfort zone.

And there is something else important, too. With the majority of environmental design, you actually have to go and visit the work in order to experience it. It's just not possible to transport it like a book or copy it like computer data. The fact that it has that element of site-specificity might make it comparable to architecture or installation.

In the world of Japanese tea ceremony there is a saying, "Ichigo Ichie" (Once in a lifetime). What it means is that "host and guest must approach every opportunity with absolute sincerity, as though it were only to happen once in a lifetime". I find it is this kind of thrilling, one-off interaction with the "guest" that really sets environmental graphic design apart from other fields. That is where its real attraction lies.

Yutaka Maeda
ujidesign, Japan

我发现环境图形设计有特别的吸引力。首先，环境图形设计的规模十分巨大。我做过很多与平面图形设计相关的项目，在实际规模上，环境图形设计是最大的。大规模意味着你在选择设计元素时有更多的选择。毕竟，一个超大的空间的装饰可以是一张照片，也可以是1,000张照片。

相反，如果在图书或屏幕上进行设计，媒介自身的尺寸就限制了你所使用的设计元素。在环境图形设计中，你无需考虑这些，可以自由地利用空间。这样一来，你可以将情绪设计变得更加情绪化，也可以把细节设计变得更加精细。一切元素在环境图形设计中都能得到加倍的体现。

当然，过多的选择也会带来问题。在图书或宣传册等小规模设计中，你可以在屏幕上尽情体验比实际尺寸更大或更小的感觉；但是环境图形设计就不能。在环境图形设计中，你只能打印出全尺寸设计的一部分，由此推断预期的整体效果。当然，你也可以利用投影仪将项目全尺寸投射在墙面上。重点是，环境图形设计需要进行更多的测试。

与纸面设计不同，环境图形设计可选用的材料也更多。每当你使用一种新材料，就必须对新理论、布局、色彩、字体大小等进行测试。

你设计的新图形必须完美无瑕，不能有丝毫令人困惑的感觉，这也是必须进行测试的原因。你在项目中所投入的时间越多，当项目完成时你的幸福感就越高。

同时，全面测试还能带来额外的新发现。因此，我认为环境图形设计能鼓励创新，将你拉出自己的"舒适区"。

更重要的是，在大多数环境设计中，你必须真正前往设计的所在地亲身体验。这不是运输一本书或复制电脑数据，这种现场特殊性让环境图形设计类似于建筑设计或装置设计。

日本茶道有一个词叫"一期一会"，意思是，"主客二人必须真诚对待每一次相聚的机会，就像对待一生中仅有一次的机会一样。"我觉得这种震撼的一次性互动与环境图形设计十分相似，也使它与其他设计领域从本质上区分开来。这正是它真正的魅力所在。

前田丰
ujidesign 设计公司，日本

CONTENTS

目录

002   Preface   前言
006   Industry   工业
022   Public-benefit   公益
032   Science   科技
052   Commercial   商业
136   Culture   文化
190   Art   艺术
238   Index   索引

London Luton Airport

Design agency: ico Design
Country: UK

Following the approval of a major development at LLA, the designers were approached to create a brand that would redefine Luton airport in the London market and inform its future direction as a passenger-focused airport. Working closely with the key stakeholders led to four core values that will inform all aspects of the brand. These are expressed in the simplicity and dynamism of the new visual identity which is a clear statement of intent of the airport's bold ambition for the future.

伦敦卢顿机场

随着伦敦卢顿机场的发展,设计师受邀为其打造一套品牌识别系统,重新定义卢顿机场在伦敦市场的定位,通过良好的导视系统将其打造为以旅客为中心的人性化机场。设计师与主要利益相关者的紧密合作为品牌奠定了四个核心价值,它们在简洁动感的视觉识别设计中得到了体现,充分表达了机场未来的雄心壮志。

设计机构:ico 设计公司 国家:英国

ABCDE56789£€$¥#@

LLA Light

ABCDE56789£€$¥#@

LLA Regular

ABCDE56789£€$¥#@

LLA Bold

ABCDE56789£€$¥#@

LLA Black

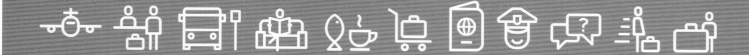

Industry

Industry

Deeley Exhibition: Made in America

Design agency: EDG Experience Design Group
Designer: Nicko Chang, Barry Marshall, Marie-Claire Hill, Karen Sorensen, Patty Kantymir Harsch
Photographer: Perry Danforth, Ihor Pona
Client: Trev Deeley Motorcycles, Harley-Davidson Canada
Country: Canada

The third Deeley Motorcycle Exhibition featured the American motorcycling industry between 1894 and 1954, focused largely on motorcycle racing, which helped keep the industry alive. Built on an earlier exhibit master plan, the visually exciting and bold graphics support feature motorcycles with unique and engaging stories – achieving a high level of visitor engagement.

迪利展览：美国制造

第三个迪利摩托展览以1894年至1954年的美国摩托制造业为主题，主要聚焦于帮助该产业保持活力的摩托车赛。以之前的展览规划为基础的视觉设计强烈而大胆，融入了摩托车的独特故事，为参观者提供了高层次的参观体验。

设计机构：EDG设计公司 设计师：尼可·张、巴里·马歇尔、玛丽－凯莉·希尔、凯伦·索伦森、帕蒂·康泰米尔·哈施 摄影：佩里·丹佛斯、伊戈尔·波纳 委托方：崔佛迪利摩托公司、哈雷戴维森加拿大公司 国家：加拿大

Industry

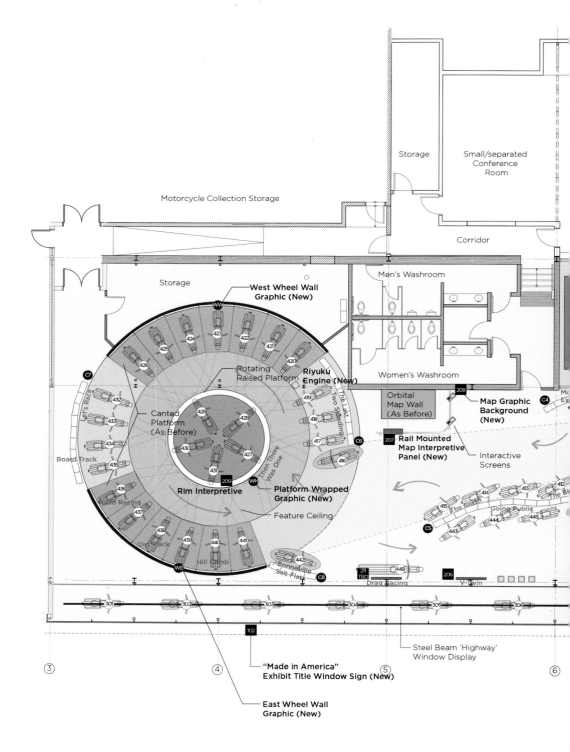

DEELEY MOTORCYCLE EXHIBITION "MADE IN AMERICA" EXHIBIT PLAN
1.0 SCALE: 1/16" = 1'-0"

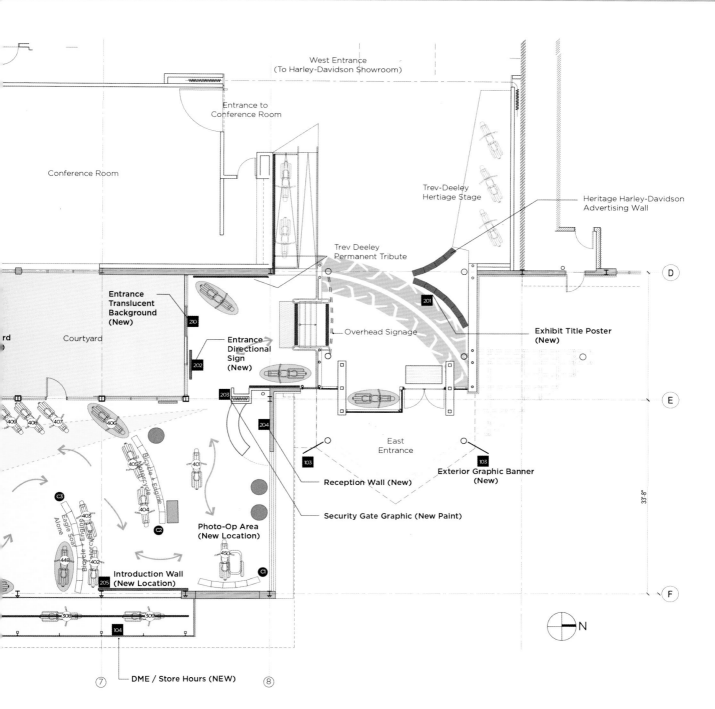

Audiovisual MAC

Design agency: KXdesigners
Designer: Katia R.Glossmann & Xavier Tutó
Graphic identity: Grafica
Photographer: KXdesigners
Client: Consorci del Mercat Audiovisual de Catalunya
Country: Spain

>>

The enclosure of the old textile factory Roca Umbert was transformed by two days to host the International Technology Market and Audiovisual Contents. The proposal was thought from the position of vehicles in the parking, to the signage to dress all the space. A complete intervention with the use of standard trade fair elements and materials, as panels, carpets, fabrics and vinyl to differentiate the enclosure areas and to create a homogeneous exhibition.

MAC 音像展览

这家旧纺织厂在两天内就被改造成了举办国际技术市场与音像内容的展览空间。设计方案考虑了从停车到所有空间的导视标识内所有环境图形因素。展板、地毯、纺织面料等标准化会展元素和材料的使用让展览变得更加统一化。

设计机构：KXdesigners 设计公司 设计师：卡蒂亚·R·格劳斯曼、泽维尔·图特 图形识别：Grafica 设计公司 摄影：KXdesigners 设计公司 委托方：加泰罗尼亚音像市场联盟 国家：西班牙

Industry

Avinguda d'Enric Prat de la Riba

La Troca

Lluís Companys

Mare de Déu de Montserrat

WC

ENTRANCE

CTUG

ACCESS HALLS A – B
TRADEFAIR'S OUTDOOR TERRACE

VIPS ROOM

| C01 | C04 | C07 | C12 | C15 | C18 | C22 | | J04 | J07 | J10 | J14 | J18 | J21 | J25 | J28 | J32 | J35 | J39 | J42 | J45 | J48 | J51 |

B08
B04

G07 | H04 | H08 | | I06
G01 | F04 | F08 | | I01

D11
D05
D01

A01

| K04 | K07 | K12 | K15 | K18 | K21 | K25 | K28 | K33 | K36 | K39 | K42 | K46 | K49 |

PRESS ROOM

| E01 | E05 | E08 | E12 | E22 | E26 |

COFFEE BAR
VIEWING A LA CARTE

PRESS CONFERENCES
COFFEE BREAK
OUTDOOR TERRACE

TRADEFAIR ENTRANCE

Industry

Xenergy Stand

Design agency: CCRZ + Decoma Design
Photographer: CCRZ
Client: Dow Chemicals
Country: Italy

>>

The exhibition design project originates from the need to promote the new building insulation product Xenergy by Dow Chemicals. The big X dominating the exhibition space is taken up by the logo and press campaign, which simultaneously appears in trade fairs and in architecture and construction industry press. It acts both as a light fixture and as a highly recognizable scenic element. The colour, the same shade of grey as the product itself, is used in order to confer a "single material" appearance to the exhibition design.

Xenergy 展台

该展览设计项目为陶氏化学新型建筑装置产品Xenergy进行宣传推广。巨大的X造型主宰了整个展览空间，遍及产品标识和宣传活动，同时还出现在展会和建筑建造业媒体上。它既是一个照明装置，又是一个具有高辨识度的场景元素。与产品同色调的灰色为展览设计外观带来了统一的材质感。

设计机构：CCRZ + Decoma 设计公司 摄影：CCRZ
委托方：陶氏化学 国家：意大利

Escort : Safety Wayfinding Signage Design

Design agency: VID Lab
Designer: Tingyi S. Lin, I-Chen Huang, Pei-Hsuan Kuan
Photographer: Tingyi S. Lin, I-Chen Huang, Pei-Hsuan Kuan
Client: Taipei City Government
Country: China

"Escort: Safety Wayfinding Signage Design Exhibition" is a partial result from the Shelters and Materials Management Project, World Design Capital Taipei 2016: Adaptive City effort. It focuses on the creation of a concise, uniform and aesthetically pleasing and informative visual signage design system. The wide public distribution of aesthetic and functional design visuals will inform and train people to immediately recognise signage information and assist them to respond to different emergency situations with reduced confusion, anxiety and stress.

护卫队：安全导视设计

"护卫队：安全导视设计展"是"世界设计之都台北2016：适应性城市"避难所与物资管理项目的一部分。它聚焦于打造精确、统一、美观、实用的导视设计系统。通过向公众大范围介绍美观实用的设计图形，他们可以快速识别导视信息并将其运用到紧急情况之中，从而减少困惑、焦虑和压力感。

设计机构：VID实验室 设计师：林廷毅、黄依晨、关佩萱 委托方：台北市政府 国家：中国

Public-benefit

奶粉
Milk Powder

泡麵
Instant Noodles

乾糧
Biscuit

飲水機
Water Dispenser

Safety Wayfinding
Signage Design Exhibition

Goods & Services
Food

礦泉水
Mineral Spring Water

罐頭
Can

沖泡飲品
Cereal & Teabag

能量食品
Energy Bar

上衣
Clothes

褲子
Pants

免洗內褲
Disposable Underpants

拖鞋
Slippers

Safety Wayfinding
Signage Design Exhibition

Goods & Services
Clothes

外套
Jacket

毛毯
Blanket

男生
Man

女生
Woman

小孩
Child

嬰兒
Baby

Safety Wayfinding
Signage Design Exhibition

Goods & Services
People

老人
Older

家庭
Family

殘障
Cripple

孕婦
Pregnant Woman

帳篷
Tent

睡墊
Sleeping Pad

睡袋
Sleep Bag

毛毯
Blanket

Safety Wayfinding
Signage Design Exhibition

Goods & Services
Bedding Sets

生活用品類
Daily Supplies

生活用品類包含盥洗用具、衛生用品及其他民生用品。

盥洗用具：
牙膏、牙刷、臉盆、毛巾、刮鬍刀、梳子等。

衛生用品：
衛生紙、衛生棉、成人及兒童紙尿褲等。

民生用品類：
考量環保概念及因應緊急狀況，提供環保餐具、紙袋、雨衣、手電筒、電池等。

Daily supplies include toiletries, hygiene items and other daily living supplies.

Toiletries:
toothpastes, toothbrushes, wash basins, towels, razors, hair brushes and so on.

Hygiene items:
toilet papers, sanitary napkins, adults and children diapers and so on.

Daily living supplies:
environmentally-friendly tableware, paper and plastic bags, raincoats, flashlights, batteries and many others are provided in consideration of environmental protection and emergent cases.

手電筒&電池
Flashlight & Battery

Public-benefit

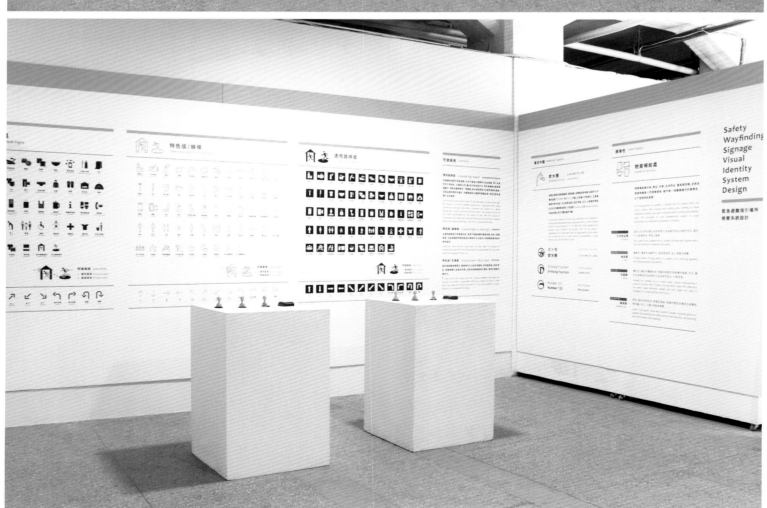

臉盆 Towel
 牙膏&牙刷 Tooth Brush & Tooth Paste
 盥洗用品 Soap
 盥洗用品 Soap & Shampoo
 盥洗室 Shower
 物資補給處 Goods & Materials
 寢室 Bedroom
 通訊處 Communication Station

 衛生棉 Pads
 衛生棉 Pads
 衛生紙 Tissue
 衛生紙 Tissue
 志工休息室 Volunteers Restroom
 緊急救護站 First Aid
 緊急救護站 First Aid
 避難管理處 Information

 毛巾 Towel
 刮鬍刀 Razor
 梳子 Comb
 塑膠袋 Plastic Bag
 餐廳 Restaurant
 垃圾分類處 Refuse Classification
 一般垃圾 Trash
 廚餘 Food Waste

 尿布 Dipper
 雨衣 Raincoat
 雨衣 Raincoat
 紙袋 Paper Bag
 塑膠 Plastic
 玻璃 Glass
 鋁箔包 Drink Cartons
 廢紙 Paper
 鐵鋁罐 Tin and Aluminum Cans

 手電筒&電池 Flashlight & Battery
 手電筒&電池 Flashlight & Battery
 環保餐具 Environment-friendly Tableware
 環保杯 Environment-friendly Cup
 衣服類 Clothes
 寢具類 Bed
 生活用品類 Supplies
食品類 Food

Public-benefit

Good Job! 2012, 2013

Design agency: UMA/design farm
Art director: Yuma Harada
Designer: Midori Hirota
Photographer: Yoshiro Masuda
Client: Tanpopo-no-ye Foundation
Country: Japan

This is an exhibition that introduces a creation of new job for handicapped people through art and design. UMA did logo design, print design, and sign of exhibition space. "Good Job!" contains not only products but also process of production, so the designers designed logo from "∞ (infinity mark)" for expressing infinity possibility and new job, social circulation.

好工作 2012~2013

本展览介绍了一系列残疾人可以通过艺术和设计所完成的创意性工作。UMA 进行了标识设计、印刷品设计和展览空间的导视设计。"好工作"展览不仅展示成品，还介绍产品的制作过程，因此设计师设计了 ∞（无限符号）来表现无限的可能、新工作和社会流动。

设计机构：UMA/ 设计农场 艺术总监：广太美岛绿 设计师：广田绿 摄影：增田喜郎 委托方：Tanpopo-no-ye 基金会 国家：日本

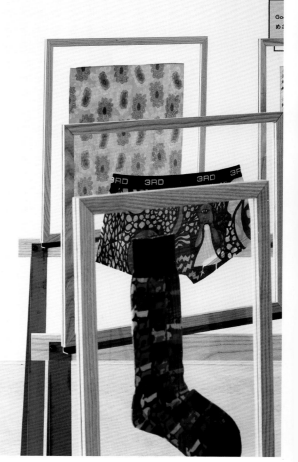

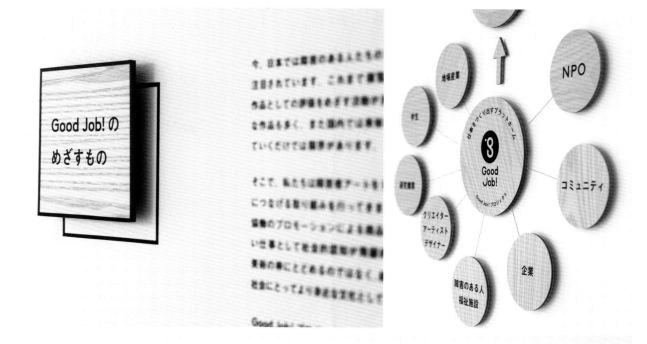

 Good Job! プロジェクト 事業内容

 Good Job!

 ① 人材育成事業

 ② ソーシャルビジネス・サポート事業

 ③ 見本市事業

 ④ アワード事業

 ⑤ 情報発信・交流事業

Public-benefit

Information Age

Design agency: Bibliothèque design
Client: Science Museum
Country: UK

Information Age is the Science Museum's biggest and most ambitious gallery to date. More than 200 years of innovation in communication and information technologies are celebrated across six zones. Each 'network' represents a different information and communication technology: Cable, Exchange, Broadcast, Constellation, Web and Cell – six networks that changed our world.

信息时代

"信息时代"是科学博物馆迄今为止最大的展览厅。200多年的信息传播技术创新被汇集在六个区域。每个"网络"展示了一种不同的信息传播技术：电缆、交换电波、广播、群集、网络和手机——这六个网络改变了我们的生活。

设计机构：Bibliothèque 设计公司　委托方：科学博物馆　国家：英国

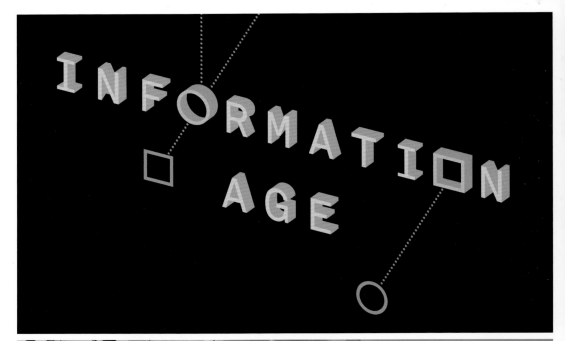

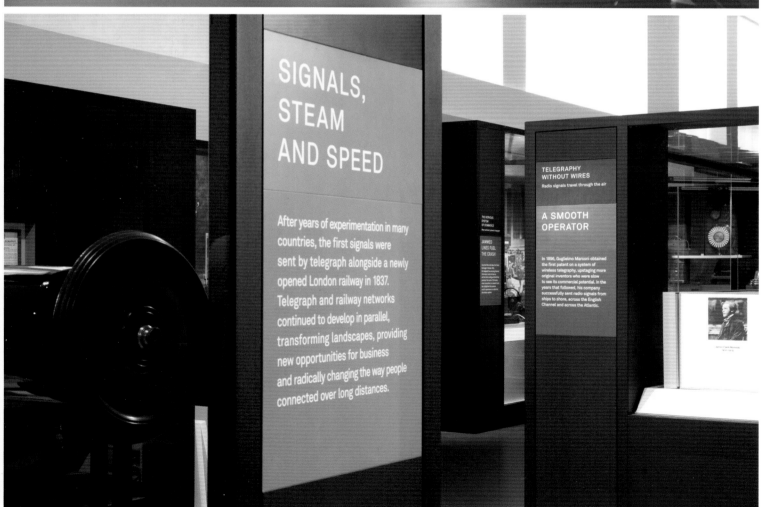

Science

Sony Wonder Technology Lab

Design agency: Lee H. Skolnick Architecture + Design Partnership
Client: Sony Corporation's Manhattan Headquarters
Country: USA

At the LHSA+DP redesigned Sony Wonder Technology Lab, visitors log in to a four story experience located at Sony Corporation's Manhattan headquarters. The experience begins when visitors join a digital community by creating profiles that link to FeliCa cards. Translated into pulses of colourful light, the profiles journey with visitors as they explore the Lab's seamlessly integrated spaces, whose continuous architecture of folded forms reflects technology's infusion in our lives. Using their FeliCa cards, visitors personalise interactions that blur boundaries between virtual and actual worlds. Whether programming robots, performing virtual surgery or sending manipulated portraits as signals, visitors creatively network through technologies that increasingly connect us.

索尼科技馆

LHSA+DP 对索尼科技馆进行了重新设计，让参观者可以在索尼公司曼哈顿总部体验四种不同的经历。参观者首先通过创造与 FeliCa 卡相关联的档案加入一个数码社区。档案被转化成彩色光脉冲，参观者随后就可以开始在科技馆中的奇妙体验了。折叠形的连续建筑形式反映了技术在我们生活中的融合。FeliCa 卡能够实现参观者的个性化互动，模糊了虚拟与现实世界的界限。无论是给机器人编程、执行虚拟调查，还是发送操作肖像信号，参观者都可以通过科技实现创意的传播与交流。

设计机构：LHSA+DP 建筑设计事务所　委托方：索尼公司曼哈顿总部　国家：美国

Science

Visions of the Universe

Design agency: Bibliothèque design
Client: Royal Museums Greenwich
Country: UK

>>

Wander through beautiful galaxies, spectacular nebulae and millions of shimmering stars in this breath-taking collection of some of the most incredible images of our universe ever made. 'Visions of the Universe' shows how man has captured images of the heavens over the centuries, from the earliest hand-drawings to photographs taken by the Hubble Space Telescope. Bibliotheque designed this exhibition for the National Maritime Museum in London, utilising large format projections and light boxes, to help reflect the range of colourful palettes that astronomers use to bring their pictures to life. It is one thing to look at these pictures in a book or on the Nasa website, but this exhibition shows them at the scale they deserve. The exhibition was split into 5 sections, following the themes of Moon, Sun, Planets, Deep Space and The Future. Bibliothèque worked with Kin Design on the interactives, including a highlight of the exhibition, the 13-metre long Mars Window. A panoramic projection of the Martian landscape as seen by NASA's Spirit, opportunity and curiosity rovers on the surface of the red planet.

宇宙景象

在美丽的银河、壮观的星云和无数闪烁的星星中漫步，该展览汇集了宇宙中最令人惊叹的图片。"宇宙景象展"展示了人类是如何捕获这些天堂图片的，从最早的手绘到现代用哈勃太空望远镜拍摄的照片。Bibiotheque设计公司为伦敦国家海洋博物馆设计的这次展览采用了大尺寸投影和灯箱来反映天文学家为我们带来的多彩宇宙景象。从图书或美国航天局网站上看这些图片与亲临现场是完全不同的，这次展览的规模成功地展现了宇宙的广袤。整个展览分为五个部分：月球、太阳、星星、外太空和未来。Bibliothèque 与 Kin 设计公司共同开发了交互式设计，其中包括一个13米长的火星窗，展示了由美国宇航局勇气号火星探测器所带回来的火星景观的全景投影。

设计机构：Bibliothèque 设计公司 委托方：格林威治皇家博物馆 国家：英国

Visions of the Moon

The Moon is one of the most recognizable and closely studied objects in the Universe. It is the only body apart from the Earth where human beings have set foot.

Its diameter is just under one third of the Earth's and it is the fifth largest natural satellite in the Solar System. At around 380,000 km (about 236,000 miles) from Earth, the Moon's light takes 1.2 seconds to reach us.

As our nearest neighbour in space, the Moon was the first astronomical object to be studied in detail through a telescope and the first to be photographed.

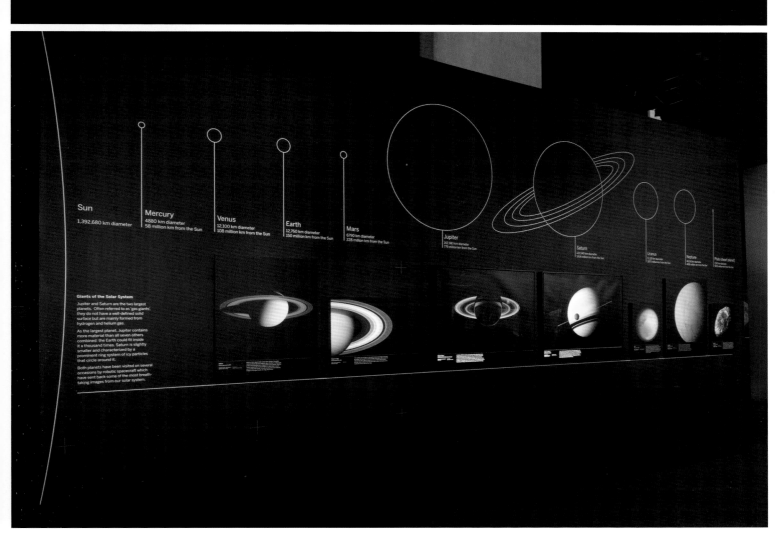

This is ISS, go ahead

Design agency: ujidesign
Designer: Ryuji Nakamura
Client: National Museum of Emerging Science and Innovation (Miraikan)
Country: Japan

This project is the permanent exhibition 'This is International Space Station (ISS), go ahead' of the National Museum of Emerging Science and Innovation, MIRAIKAN. Signage is fully worked out to explain the space station as well as provide advanced scientific information to visitors as lucidly as possible. One of the booth explaining what environment prerequisites are needed for human beings to exist. The columnar wall is an exhibition of a quasi-Earth designed not only to be experienced by sight, but also sound and the sense of touch. Other exhibitions also examine the various means of expressions with the aim of both enlightening visitors and making the spaces comfortable.

这里是国际空间站，前进

本项目是日本未来科学馆的"这里是国际空间站，前进"展览。导视设计既能全面解读空间站，又能为参观者提供清晰明确的先进科技信息。其中一个展亭说明了人类生存的先决条件，柱状墙壁被设计成类似地球的展览，不仅提供视觉体验，还能提供声觉和触觉体验。其他展览空间的设计同样兼顾了为参观者提供信息和保证参观者的舒适感两方面。

设计机构：ujidesign 设计公司　设计师：中村龙治
委托方：日本未来科学馆　国家：日本

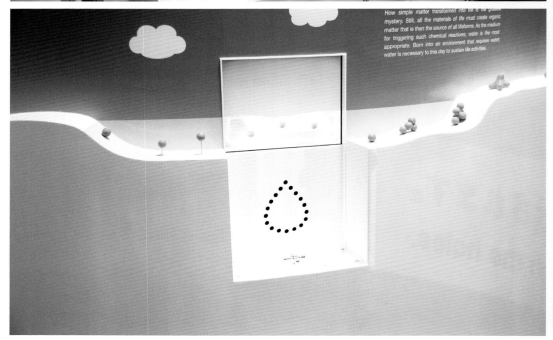

Science

Special Exhibition: The End of the World - 73 Questions We Must Answer

Design agency: ujidesign
Designer: Takeshi Tanio
Photographer: Takumi Ota
Client: National Museum of Emerging Science and Innovation (Miraikan)
Country: Japan

One year after the Great East Japan earthquake (March 11, 2011), Special Exhibition "The End of the World - 73 Questions We Must Answer" was held in the National Museum of Emerging Science and Innovation (Miraikan). Nuclear power plant was exploded and people faced crisis that radioactive materials poured on the ground. The exhibition asks the role of the science technology again, and finds new story of hopes which begins from end. This is the graphic design and sign design for it. The illustration of Takashi Taima is the visual identity of this exhibition. The designers developed it from exhibition space to signage and publicity. Although there are serious themes like death, disaster, disease, the illustration attracts people's wonder and main colour yellow makes warm feeling. To show the real data, the designers used pictograms, diagrams and illustration. It communicates visitors visually and helps them easy to understand.

特殊展览：世界的尽头——73个我们必须回答的问题

在东日本大地震（2011年3月11日）一周年纪念之时，"世界的尽头——73个我们必须回答的问题"特殊展览在日本未来科技馆进行展出。核电站爆炸令人们面临着放射性原料流向地面的危机。展览重新审视了科技的角色，希望从尽头处找到新希望。ujidesign为展览提供了平面图形设计和标识设计。当麻隆的插画被用于展览的视觉识别设计。设计师在展览空间、导视设计和公共宣传都使用了插画。尽管展览有死亡、灾难、疾病等严肃的主题，插画能吸引人们的好奇，而黄色也带来了温馨的感觉。为了展示真实数据，设计师使用了象形图、图表和插图进行说明。它们从视觉上进行传播，有助于参观者的理解。

设计机构：ujidesign设计公司　设计师：谷尾武
摄影：太田拓实　委托方：日本未来科学馆　国家：日本

Science

Science

Sea Change: Boston

Design agency: Sasaki Associates
Country: USA

It is projected that sea levels will rise two feet by mid-century and six feet by 2100. The new tide line will transform the coastal landscape of Greater Boston and increase the probability of a major storm devastating the metropolitan region. The Sea Change: Boston exhibition at District Hall examines Boston's vulnerabilities to sea level rise and demonstrates proactive design strategies at the building, city, and regional scale. The exhibition is intended to catalyze conversations with a broader audience about the tough questions and regional implications of sea level rise. Sea Change: Boston not only proposes smaller scale design interventions, but also zooms out to understand the regional implications of sea level rise in order to proactively design and plan for the Greater Boston area. The exhibition is designed to communicate the complexities of sea level rise and resilience in a clear, accessible, and balanced way, with the goal of raising awareness among the general public and encouraging call for action at the city and regional scales.

海洋变化：波士顿

据预测，海平面到 2050 年将上升 2 英尺，到 2100 年将上升 6 英尺。新的低潮线将改变大波士顿地区的沿海景观并增加毁灭性的暴风发生的几率。"海洋变化：波士顿"展览审视了波士顿面对海平面上升的脆弱性，同时还演示了前瞻性建筑、城市、区域设计策略。展览旨在促使受众针对这些严峻的问题进行更广泛的对话。"海洋变化：波士顿"展览不仅提出了小规模的设计措施，还上升到了整个区域的高度，提议对大波士顿地区进行前瞻性设计和规划。展览设计通过清晰、简单、均衡的方式表现了海平面上升和恢复的复杂性，力求提升普通民众的认识，呼吁在城市及区域范围内采取行动。

设计机构：Sasaki 事务所　国家：美国

MAC

Design agency: KXdesigners
Designer: Katia R.Glossmann & Xavier Tutó
Graphic identity: Grafica
Photographer: KXdesigners
Client: Consorci del Mercat Audiovisual de Catalunya
Country: Spain

The Old textile factory Roca Umbert has lodged the 8th edition of the MAC, the international technology and audiovisual contents market. The intervention had required of a total transformation and adaptation to satisfy the necessities of the project. The space has allowed an approach as in an exhibition. The proposal details are the use of standard trade fair elements, the treatment of the light, and the use of MAC's identity colours.

加泰罗尼亚音像市场联盟展览会

罗卡乌伯特纺织厂举办了第八届加泰罗尼亚音像市场联盟（MAC）展览会。项目设计要求对厂房进行全面的改造，以适应展览的需求。设计方案的细节包括标准展会元素、灯光的处理以及MAC的标志性色彩的应用。

设计机构：KXdesigners设计公司　设计师：卡蒂亚·R·格劳斯曼、泽维尔·图特　图形识别：Grafica设计公司　摄影：KXdesigners设计公司　委托方：加泰罗尼亚音像市场联盟　国家：西班牙

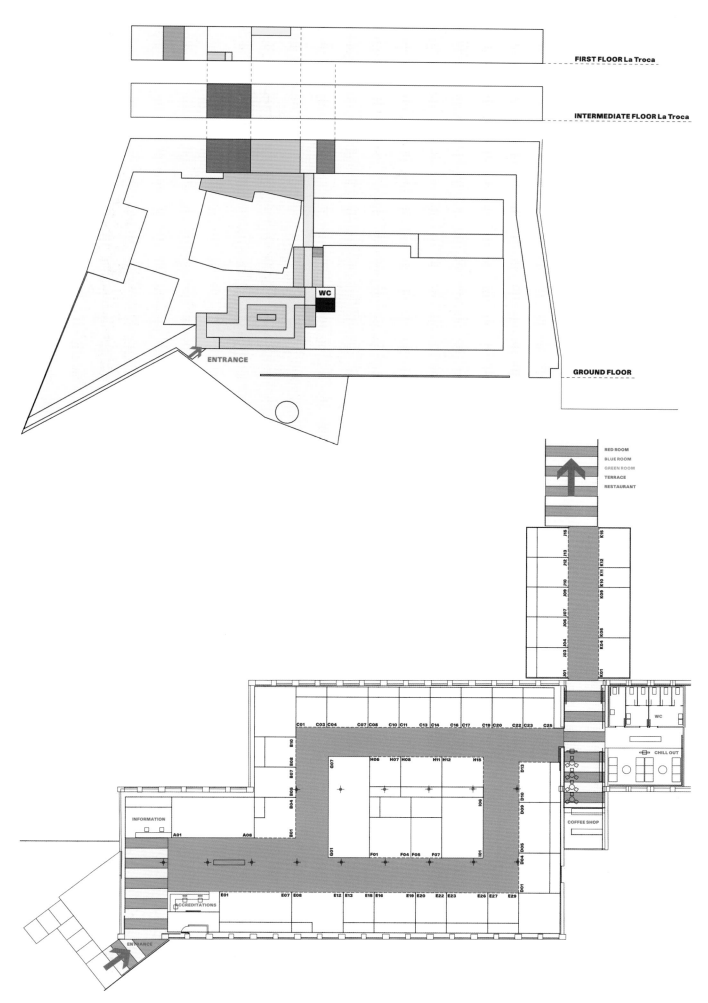
Science

La Casa del III Millennio

Design agency: Frits van Dongen - CIE with D'Apostrophe
Designer: Donatello D'Angelo, Cosimo Damiano D'Aprile
Photographer: D'Apostrophe
Client: CNA - Artigianato Pratese
Country: Italy

The exhibition "La Casa del III Millennio", held in Prato, has been the result of a research path on the "house system". The aim was that of drive the creatives in finding alternative solutions both for the envelope and the components of the house. The exhibition's fitting and visual system concept has been developed following the idea of sustainability and materials recycle. So was born the idea of an installation composed by a T-shirt on the facade of Palazzo Pacchiani, Historical Prato building housing the event, or that of create the intern of an oneiric cathedral whose space has been defined by arcades and cypress columns. The catalog, as well as the whole fitting visual system, have been realised through a visual system similar to the idea of the future artifact and the natural geometry, emphasised by the employment of a chromatic palette expressly composed, and by materials of contemporaneous workmanship but sustainable.

三千年之家

"三千年之家"展览在意大利普拉托展出，是对"住房系统"的深刻研究展示。展览的目标是寻找住宅外墙和内部组件的创新解决方案。展览装置和视觉系统概念遵循了可持续设计和材料回收的概念。因此，设计师为展览的举办地帕奇亚尼宫穿上了"新衣"，并且通过拱廊和柏树打造了临时的教堂。展览目录和整套视觉识别系统都与手工艺品和自然几何图形相关联。固定的色彩搭配突出了设计，各种手工艺材料全部具有可持续特征。

设计机构：Frits van Dongen 设计公司、D'Apostrophe 设计公司 设计师：多纳泰罗·德安杰洛、科西莫·达米亚诺·达普利勒 摄影：D'Apostrophe 设计公司 委托方：CNA——普拉托手工业协会 国家：意大利

KI Furniture

Design agency: Bailey Lauerman
Creative director: Ron Sack
Designer: Brandon Oltman
Client: KI Furniture
Country: USA

KI Furniture showroom treatments. In 2012, KI Furniture launched Lightline, a moveable glass wall system dedicated to the enhancement of light. The "Lightline" type treatment was molded out of white acrylic and suspended in the centre of the product (four glass walls). The whole signage system is directly applied on the interior wall. The extruded effect of 3D letters and shadows in special light make the signs striking even in a long distance.

KI 家具展

本项目为KI家具样板间陈列设计。2012年，KI家具推出了Lightline系列移动玻璃墙系统，专为改善室内光线条件。Lightline系列产品的展示标识采用了白色亚克力材料，悬挂在四层玻璃墙中间。整个标识系统直接应用在室内的墙面上，标识中的字母采用立体的凸出效果，特殊的灯光打在字母上所产生的阴影，即使在远处都非常的引人注目。

设计机构：Bailey Lauerman 设计公司 创意总监：罗恩·萨克 设计师：布兰登·欧特曼 委托方：KI家具 国家：美国

3M CTC Innovation Facility

Design agency: THERE
Designer: Paul Tabouré, Jon Zhu
Photographer: Richard Glover
Client: Colliers Project Services
Country: Australia

Following the successful completion of an extensive branded workplace environment program for 3M's Australian HQ, THERE were asked to partner with the team at Colliers Project Services again, to brand a tactile and interactive 3M innovation space that would create links between the different technology divisions within the business. The CTC Innovation Facility, opened by Greg Combet, the Minister for Innovation and Industry, is an opportunity for the different 3M technology divisions to display their innovative developments on custom modular stands. The space features multi-touch screen displays, media walls and LCD screens that demonstrate the diverse 3M technologies in unique and interesting ways. As the primary technology showcase area of 3M, the space needed to have a sense of cumulated achievement, taking the 'trophy room' concept to the next level. The branded space provides an atmosphere of collaboration that enhance 3M's relationship with their customers and partners, increasing their understanding of one another, and ultimately informing positive sales and profit growth.

3M CTC 创意馆

在帮助3M公司澳大利亚总部完成办公环境的品牌化设计之后，THERE设计公司受委托与Colliers项目服务公司共同为3M公司打造一个创意空间，将该公司不同的技术部门联系起来。CTC创意馆由创新与工业部部长格雷格·康贝特创办，是3M公司不同的技术部门展示自己创新发展的平台。空间以多点触摸屏展示、媒体墙和液晶显示屏为主，以独特、有趣的方式展示了各种各样的3M技术。作为3M公司的主要技术展示区，这一空间需要一种累积的成就感，将"奖杯陈列室"上升到一个新的高度。这个品牌空间营造出一种能够提升3M与客户和合作者关系的合作氛围，增加了他们的相互理解，从而最终实现销售额和利润的提升。

设计机构：THERE设计公司 设计师：保罗·塔布莱、乔恩·朱 摄影：理查德·格罗夫 委托方：Colliers项目服务公司 国家：澳大利亚

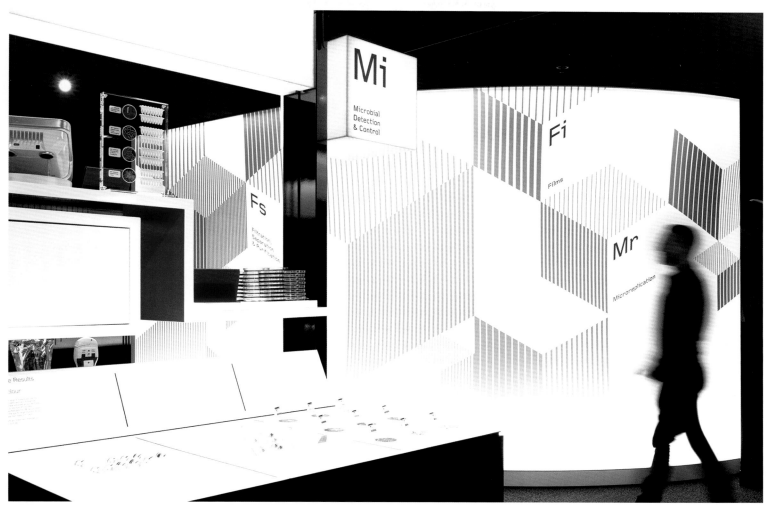

Commercial

10/10 Arper Roadshow

Design agency: THERE
Designer: Paul Tabouré, Jon Zhu
Photographer: Simon Hancock
Client: Stylecraft
Country: Australia

THERE helped Stylecraft celebrate their 10th successful year in partnership with Italian furniture supplier Arper with an engaging series of events at each of Stylecraft's five showrooms across Australia. By evolving Stylecraft's branding the designers created an event identity and circular design language that informed promotional collateral and in store display items. Each Arper product was represented by a striking colour which was integrated throughout, from external signage and interior display to banners, posters, invitations, badges and DVD packs. The events, aimed at a discerning A&D audience, had outstanding attendance and generated high media interest, raising Stylecraft's profile even higher within the industry.

10/10Arper 家具巡回展

THERE 设计公司帮助 Stylecraft 庆祝了他们与意大利家具供应商 Arper 的 10 周年合作，在 Stylecraft 公司横跨澳洲的五个展览室内进行了一系列活动和展览。为了体现 Stylecraft 的品牌形象，设计师打造了一个互动标识和圆形设计语言。每件 Arper 产品都以冲击性的色彩展示出来，这些色彩贯穿了整个场景，从外部标识到内部展览的横幅、海报、邀请函、胸卡和 DVD 包装盒。展览活动的受众是独具慧眼的视听受众，获得了出色的到场率，吸引了媒体的高度注意，进一步提升了 Stylecraft 在业内的品牌地位。

设计机构：THERE 设计公司　设计师：保罗·塔布雷、乔恩·朱　摄影：西蒙·汉考克　委托方：Stylecraft 公司　国家：澳大利亚

Commercial

Commercial

C.P. Hart Showroom Design

Design agency: I-AM London
Creative director: Jon Blakeney, Pete Champion
Designer: James Coates
Client: C.P. Hart
Country: UK

In the wake of expanding their network of UK showrooms and developing their online presence, C.P. Hart commissioned I-AM to develop their retail format strategy and showroom experience concept. I-AM worked collaboratively with C.P. Hart's senior management to define a sector-leading approach that consolidates their position as the high-end brand of choice for discerning customers and professionals. Demonstrating how physical and digital retail are becoming increasingly integrated, I-AM created a digitally augmented environment that showcases C.P. Hart's key collections in a seamless, limitless 'virtual showroom'. This new direction is being piloted at C.P. Hart's Chelsea site. C.P. Hart's core value of 'inspiration' was conceptually translated by I-AM by commissioning Jason Bruges Design to create a dramatic multimedia art installation that celebrates the beauty of flowing water. The overall result is a strong brand statement that supports C.P. Hart's goal of being 'the best luxury bathroom retailer'.

C.P. Hart 展厅设计

在对伦敦展厅和在线网站进行扩展之后，C.P. Hart 卫浴决定委托 I-AM 为他们开发一套零售策略和展厅体验概念。Hart 公司决定通过展厅设计来巩固他们高端卫浴品牌的地位。为了展示网店和实体店的相互融合，I-AM 打造了一个充满了数字化设计的展览环境，通过"虚拟展厅"展示了 C.P. Hart 的主要产品。C.P. Hart 的核心价值观"灵感"被 I-AM 转化为戏剧化的多媒体艺术装置（由 JasonBruges 设计公司设计），突出了流动的水之美。整体设计强烈地体现了 C.P. Hart 品牌"做最好的奢华卫浴零售商"的目标。

设计机构：I-AM 伦敦 创意总监：乔恩·布莱克尼、皮特·钱皮恩 设计师：詹姆斯·科茨 委托方：C.P. Hart 公司 国家：英国

THERE MUST BE QUITE A FEW THINGS THAT A HOT BATH WON'T CURE, BUT I DON'T KNOW MANY OF THEM. — Sylvia Plath

C.P. HART

I LIKE TO GIVE MY INHIBITIONS A BATH NOW AND THEN.

Oliver Reed

ROMA CONSOLE BASIN 1200MM

MODEL NO: ROWB1201WW
SIZES AVAILABLE:
1200W X 500D X160H - 1 TAP HOLE
1200W X 500D X160H - 3 TAP HOLES
MATERIAL: CERAMIC
FINISH: WHITE

£216

MOTO - SINGLE LEVER BASIN MIXER

MODEL NO: MO110DNCP
SIZES AVAILABLE:
56W X 146D X 163H. NO POP UP WASTE
56W X 169D X 163H. WITH POP UP WASTE
MATERIAL: BRASS
FINISH: CHROME

£190

Commercial

Domus Tiles Showroom

Design agency: I-AM London
Creative director: Jon Blakeney, Pete Champion
Designer: Nick Wills, Tanya Fairhurst
Photographer: Kristen McCluskie
Client: Domus
Country: UK

Over the course of a ten-year relationship with Domus, I-AM have reviewed and developed their brand and sales process and created two inspirational showrooms in Clerkenwell and London's West End. Domus has provided a unique environment for clients to explore new ideas. The design concept for the Clerkenwell showroom is 'back to front' and was inspired by visiting the Domus warehouse and their many suppliers in Northern Italy. The showroom has been returned to its foundations, stripped of artificial display walls and counters to reveal the body of this generous space. I-AM's concept for the West End showroom develops the idea that 'opposites attract'. By opening the space and using dynamic, innovative materials, the showroom is a versatile event and meeting space and a practical working environment for the Domus team. There is no fixed route in the showroom. However, the ubiquitous signage system leads visitors like a tour guide.

Domus 瓷砖展示厅

在与 Domus 公司十年的合作中，I-AM 为他们的品牌营销进行了评估和开发，并且在克勒肯维尔和伦敦西区打造了两间充满创意的展厅。Domus 公司为客户提供了一个探索新观念的独特环境。克勒肯维尔展厅的设计概念"前后倒置"的设计灵感来自于参观 Domus 公司的仓库和他们在南意大利的供应商。展厅返璞归真，摒弃了人工展示墙和柜台，充分显露了宽敞的空间。I-AM 在西区展厅的设计中开发了"对立吸引"的概念，利用动感、创新的材料打开空间，既适合举办多种活动和会议，同时也可供 Domus 公司团队进行日常办公。整个展厅并没有固定的路线，无处不在的导视系统却像导游一样引导着参观者前行。

设计机构：I-AM 伦敦　创意总监：乔恩·布莱克尼、皮特·钱皮恩　设计师：尼克·威尔斯、塔尼亚·费尔赫斯特　摄影：克里斯汀·麦克克劳斯基　委托方：Domus 公司　国家：英国

Commercial

designjunction 2014

Design agency: Bravo Charlie Mike Hotel
Photographer: John Hooper
Client: designjunction
Country: UK

designjunction showcases furniture, lighting and product design from around the world, presented against industrial backdrops. Bravo Charlie Mike Hotel were commissioned to brand and design the signage and environmental graphics for designjunction's trade shows. The 2014 London show was spread over three floors of the Old Sorting Office, a large disused Post Office building in central London. The signage and environmental graphics are designed to be large, simple and bold to help visitors navigate the large, sprawling space. The signage was printed onto Correx sheets allowing directional totems to be easily fabricated on site and signage to be fixed to the rough walls with staples reflecting the show's temporary nature in the unfinished, industrial venue.

2014 designjunction 展览

designjunction 展览展出来自于全球各地的家具、照明和产品设计。Bravo Charlie Mike Hotel 设计公司受委托对 2014 designjunction 展览进行品牌营销和标识与环境图形设计。2014 伦敦站占据了伦敦中心区一座废弃邮局大楼的老邮件分拣处的三个楼层。导视和环境图形设计大胆而简单，帮助参观者在广阔的空间里实现定位导航。导视标识被印在 Correx 板材上，让引导标志可以快速简单地在现场进行组合装配并直接通过图钉安装在粗糙的墙面上，反映了展览的临时性。

设计机构：Bravo Charlie Mike Hotel 设计公司 摄影：约翰·胡泊尔 委托方：designjunction 国家：英国

Commercial

Sistema Como 2015

Design agency: CCRZ + Studio Brambilla Orsoni
Designer: Paolo Brambilla, Eugenio Castiglioni
Photographer: CCRZ
Client: Camera di Commercio Como
Country: Italy

This design is the corporate identity for Sistema Como 2015, a territorial project that aims at attracting the attention of the Universal Exhibition visiting countries by presenting the companies, skills and excellences of the Como area.

2015 科摩系统展

本项目是为 2015 科莫系统展所进行的整体形象设计。该展览旨在通过展示科莫地区的公司、技术和优秀产业来吸引全世界更多的人到当地投资。

设计机构：CCRZ + Brambilla Orsoni 工作室 设计师：保罗•布兰比拉、欧亨尼奥•卡斯蒂廖尼 摄影：CCRZ 委托方：科莫商业局 国家：意大利

Commercial

Uniqlo

Design agency: Bravo Charlie Mike Hotel
Client: Uniqlo
Photographer: Neil Bridge
Country: UK

Bravo Charlie Mike hotel were commissioned to design the press launch of the Uniqlo AW11 collection which featured pieces with advanced technological textiles and processes. The designers created environments with simple experiments to highlight and reflect the specific features of the items of clothing. All elements within the environments from the display surfaces and experiments to the typography on the panels strictly aligned to a visible 3D grid.

优衣库

Bravo Charlie Mike Hotel设计公司受邀为优衣库AW11系列的媒体发布会进行设计，该系列服装以高新技术面料和制作过程为卖点。设计师用简单的试验来营造环境氛围，突出并反映了服装的特色。从展示平面、试验展示到文字设计，所有环境元素都与视觉三维网格严格对齐。

设计机构：Bravo Charlie Mike Hotel设计公司 委托方：优衣库 摄影：尼尔·布里奇 国家：英国

Spritmuseum

Design agency: Stockholm Design Lab
Designer: Lomar Arkitekter, Dan Arne
Client: Spirits Museum
Country: Sweden

Solution
Based on research identifying the potential for a new, well-travelled, sophisticated audience, the designers developed a brand platform, positioning and personality, and a new, much shorter name. 'Spritmuseum' conveyed the new focus on spirits and, set in Absolut's modified Futura font, it gave a nod to the main sponsor. In the join between the sheds, we saw a cocktail glass: this became a symbol for navigation, souvenirs and bags. And the simple design programme the designers developed guided the production of all the museum's communications.
Result
A reduction of 70% in the letter-count of the museum's name was appreciated by listings editors everywhere. Perhaps more importantly, the designers created a potent brand for Spritmuseum and helped to drive its complete reinvention. Annual visitor numbers immediately soared to 150,000, and very quickly, Spritmuseum became established as the newest, coolest destination among many on Djurgården. And it's still showing no sign of getting drunk on its own success.

精神博物馆

方案
在对受众人群进行调研之后，设计师为博物馆打造了品牌平台、品牌定位、个性化设计和更简短的新名字。"精神博物馆"（Spritmuseum）体现了聚焦于"精神"的新观点，通过经Absolut改良的Futura字体呈现，获得了主赞助人的认同。鸡尾酒杯形成了导航的象征，并且出现在纪念品和包装袋上。简单的设计有效地实现了博物馆在各个方面的传播。

成果
博物馆名称的字母数被减少了70%，深受各地的编辑人员欢迎。更重要的是，设计师为精神博物馆打造了一个强有力的品牌，有助于它进行全面的改造。博物馆的年度参观者迅速增加到了150,000人次，使其成为了尤尔格丹最新潮、最时尚的景点。

设计机构：Stockholm Design Lab设计公司 设计师：洛马尔·阿齐泰克特、丹·阿恩 委托方：精神博物馆 国家：瑞典

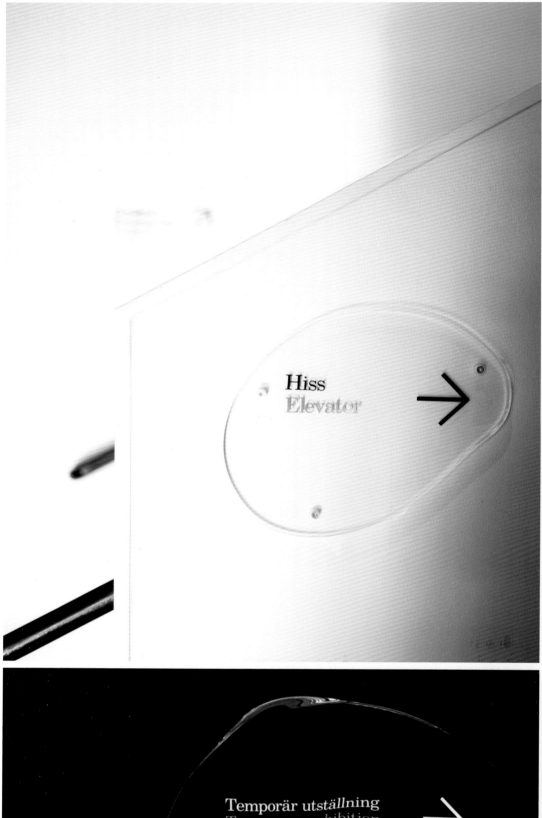

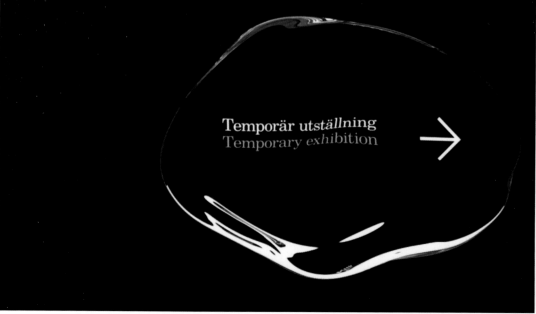

Commercial

Brunner Exhibition Stand – Orgatec 2014

Design agency: Ippolito Fleitz Group GmbH
Designer: Peter Ippolito, Gunter Fleitz, Tanja Ziegler, Katja Heinemann, Alexander Fehre, Anne-Laure Minvielle
Photographer: Andreas Körner
Client: Andreas Körner
Country: Germany

At this year's Orgatec, Brunner presents new product innovations and its portfolio of chairs and tables designed for office use. The goal was to use its innovations to present Brunner as a dynamic, transparent company, while placing the products on a comprehensible and suitable platform. Using a total of 18 kilometres of polypropylene cord, an open and transparent exhibition stand was created, where subtle colours and almost graphical wall filters form the perfect platform for Brunner's products.

Brunner 公司展位——2014科隆办公家具展

在本届 ORGATEC 科隆国际办公家具及管理设施展上，Brunner 公司展示了其特别为办公空间所设计的桌椅系列。本次展台设计的目标：借众多创新之力展示 Brunner 公司充满动力的开放企业形象，同时为其产品提供一个易于解读的恰当的展示舞台。整座展台消耗线绳总长18千米，制造出一个开放、透明的空间。简约的色彩和近乎图案化的墙面效果为 Brunner 产品提供了一个完美的展示舞台。

设计机构：Ippolito Fleitz 设计集团 设计师：彼得·依普利托、甘特·福莱兹、塔尼亚·齐格勒、卡特娅·海尼曼、亚历山大·菲尔、安妮-劳尔·敏维尔 摄影：安德里亚斯·科尔内尔 委托方：安德里亚斯·科尔内尔 国家：德国

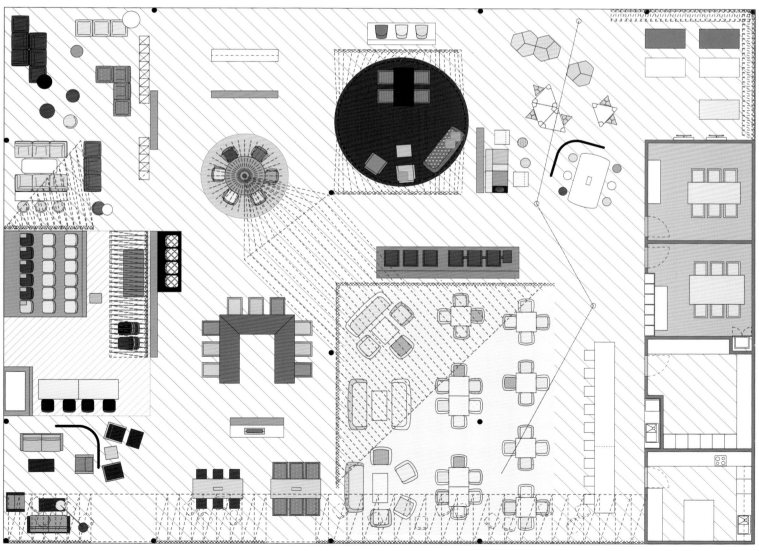

Commercial

fold chair
Design: Robert Hoffmann
Design Concept

Walter Knoll Exhibition Stand – Orgatec 2014

Design agency: Ippolito Fleitz Group GmbH
Designer: Gunter Fleitz, Peter Ippolito, Tilla Goldberg, Axel Knapp, Alexander Aßmann, Frank Peisert, Kim Angenendt, Florian Holzer, Andrea Koppenborg, Verena Schiffl
Photographer: Zooey Braun
Client: WALTER KNOLL AG & Co. KG
Country: Germany

Communication and identity are key themes of our new working worlds and provide the motto for this year's Walter Knoll presence at the Orgatec 2014. The exterior skin of the trade stand is illustrated with a graphic translation of the free flow of integrated communication – while allowing the eye to catch glimpses of the stand's interior.

Walter Knoll 公司展位——2014 科隆办公家具展

交流与认同感是全新办公理念的核心主题，也是 2014 年 ORGATEC 科隆国际办公家具及管理设施展上 WalterKnoll 公司参展的指导思想。展台外围包裹的图样以流畅联网的自如交流为主题，断断续续的幕布使展台内部的环境气氛能够些许渗透到其外，吸引访客目光。

设计机构：Ippolito Fleitz 设计集团 设计师：甘特·福莱兹、彼得·依普利托、蒂拉·哥德伯格、阿克塞尔·克纳普、亚历山大·阿伯曼、弗兰克·佩色特、吉姆·安吉奈特、弗洛莱恩·霍尔泽、安德里亚·库本伯格、维利纳·施福尔 摄影：佐伊·布朗 委托方：WALTER KNOLL 公司 国家：德国

Commercial

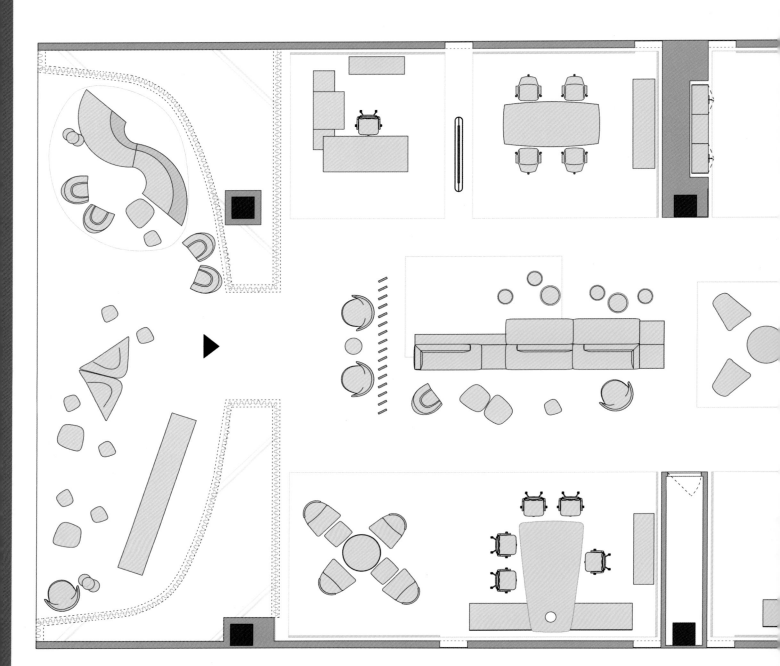

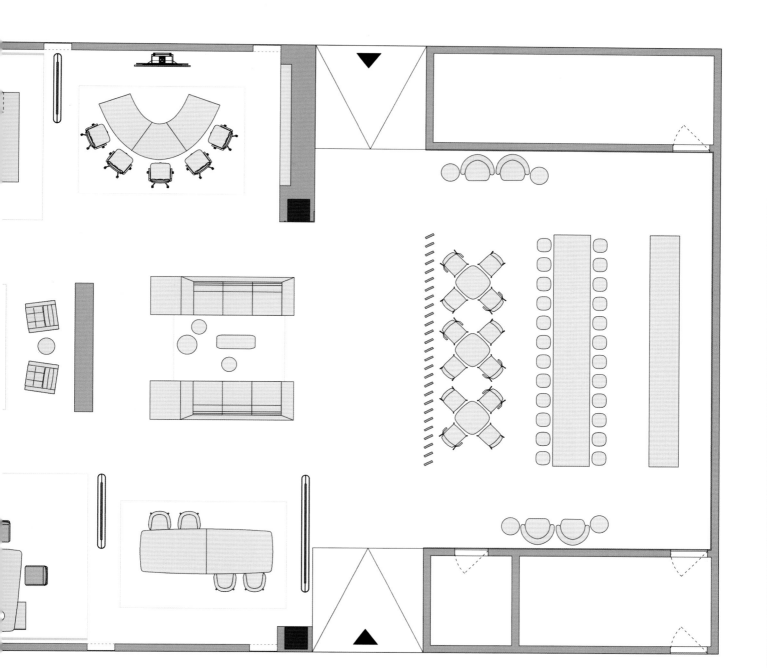

Commercial

Burkhardt Leitner EuroShop 2014

Design agency: Ippolito Fleitz Group GmbH
Designer: Peter Ippolito, Gunter Fleitz, Tilla Goldberg, Jonas Hertwig, Verena Schiffl, Axel Knapp, Caroline Lintl
Photographer: Zooey Braun
Client: Burkhardt Leitner constructiv GmbH & Co. KG
Country: Germany

The Burkhardt Leitner constructiv exhibition stand at the EuroShop 2014 showcases creative ways of employing its modular architecture systems. It also honours the company's founder on the occasion of his seventieth birthday. This year, the company's core strengths – precision, innovation and modularity – team up with a strong emotional pull. The slogan 'In Love with Detail' frames surprising new ways of employing the company's functional, minimalist system architecture, attracting attention from new target groups as well as the industry itself.

Burkhardt Leitner 公司展位——2014 欧洲商店展

Burkhardt Leitner 建材公司在本届 EuroShop 欧洲店铺产业展览会上的展位设计围绕该公司开发的模块式建筑系统，展示了其中蕴藏的无限可能性，并将聚光灯投向庆祝七十华诞的公司创始人。在展示公司以精准度、创新性和模块化为代表的核心强项的同时，另将高度的情感性融汇其中。本次展位以"In Love With Detail"即"细节之爱"为主题，运用兼备高度功能性和极简审美风格的系统化建筑手段，搭建出多种令人意想不到的应用实例，既吸引了新的目标群体，又受到资深行业人士的瞩目。

设计机构：Ippolito Fleitz 设计集团 设计师：彼得·依普利托、甘特·福莱兹、蒂拉·哥德伯格、乔纳斯·赫特维希、维利纳·施福尔、阿克塞尔·克纳普、卡洛琳·林特 摄影：佐伊·布朗 委托方：Burkhardt Leitner 建材公司 国家：德国

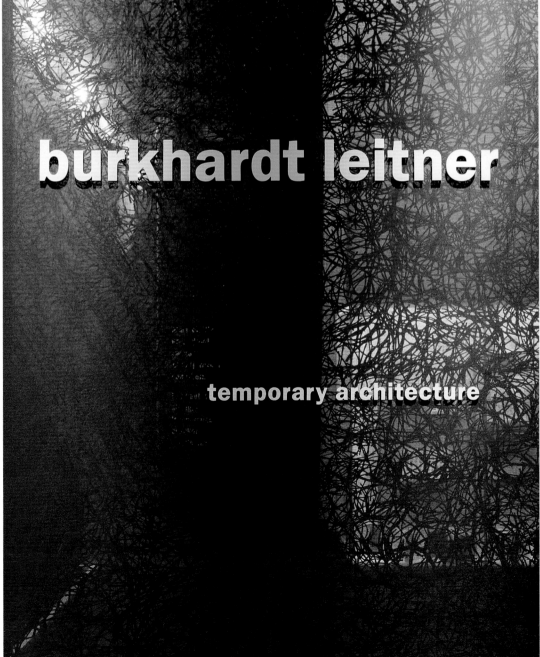

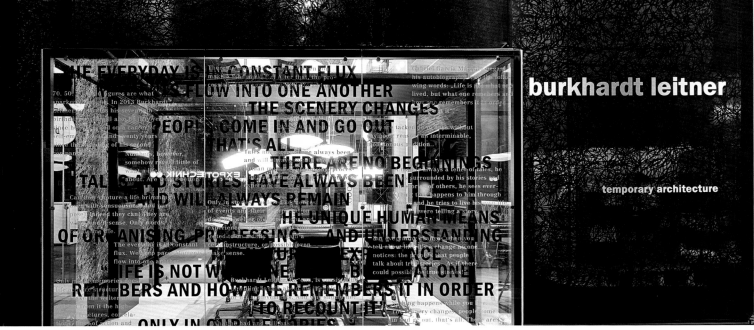

Commercial

MEXICO CITY
NEW PLYMOUTH

OSLO
PORTO
PRAHA

RABAT
RIGA
RIYADH

ST GALLEN
SAINT PETERSBURG
SAO PAULO
SEOUL
SHANGHAI
SINGAPORE
STUTTGART

TEHERAN
TEL AVIV
THESSALONIKI

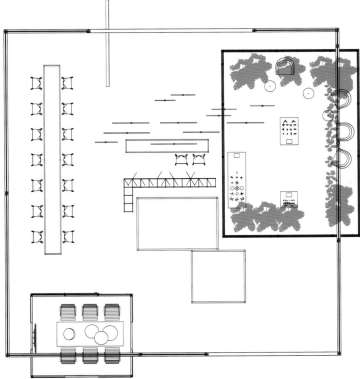

Commercial

Brunner – Fair Stand Salone Internazionale del Mobile

Design agency: Ippolito Fleitz Group GmbH
Designer: Peter Ippolito, Gunter Fleitz, Tilla Goldberg, Jörg Schmitt, Daniela Schröder, Axel Knapp, Martin Berkemeier, Frank Faßmer, Anna Maier, Leonie Beck, Monther Abuhmeidan
Photographer: Francesco Di Loreto
Client: Brunner GmbH
Country: Italy

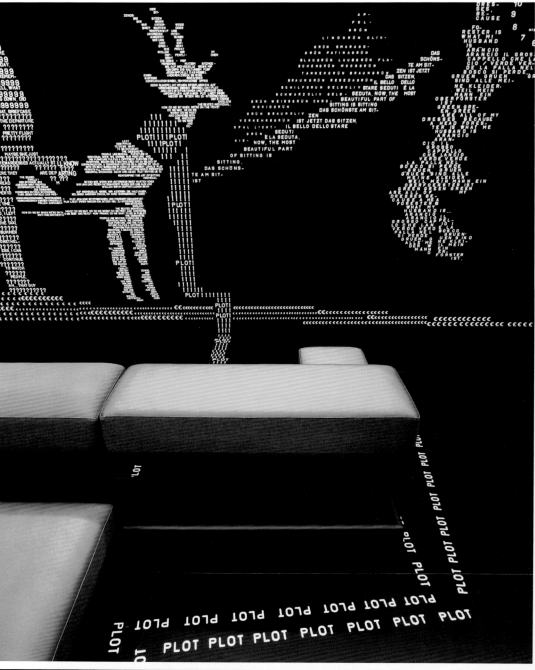

The exhibition stand creates an intense spatial experience in which to stage the innovative seating elements 'Plot' and 'Hoc'. The strictly black and white setting takes a back seat to the colourful seating objects, which invite passers-by to stop and test them. The walls reflect thoughts and inspirational ideas that arise during the contemplative act of sitting. They are completely covered in an installation of letters, forming associative figures and scenarios. At closer look, poetic phrases, dialogue sequences and questioning thoughts emerge – a reminiscence of poetry. A brightly illuminated ceiling of contoured textile fins is suspended above the seating landscape, conjuring up the image of a three-dimensional expanse of clouded sky.

Brunner 公司——国际移动沙龙展位

展位营造出一种强烈的空间体验，展示了 Plot 和 Hoc 系列创新座椅产品。黑白布景为多彩的座椅提供了良好的背景，吸引着过往行人驻足观看。墙壁上反映了座椅的设计理念和内在精神。它们整个由文字装置覆盖，形成了组合图形和场景。走进来看，诗句、对话和问话——浮现，就像诗歌一样。座椅上方悬挂着明亮的波浪形天花板，就像三维立体版的蓝天白云。

设计机构：Ippolito Fleitz 设计集团 设计师：彼得·依普利托、甘特·福莱兹、蒂拉·哥德伯格、乔治·施密特、丹妮拉·施罗德、阿克塞尔·克纳普、马丁·伯克米尔、弗兰克·法伯梅尔、安娜·迈尔、利奥尼·贝克、蒙齐尔·阿布美丹 摄影：弗朗西斯科·迪洛雷托 委托方：Brunner 公司 国家：意大利

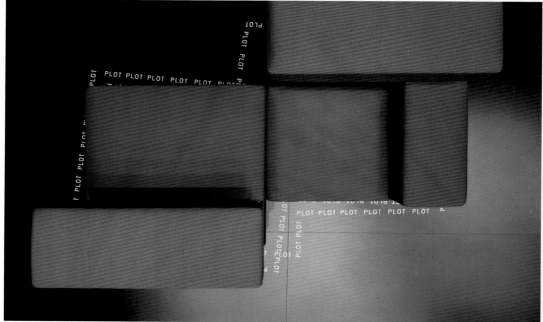

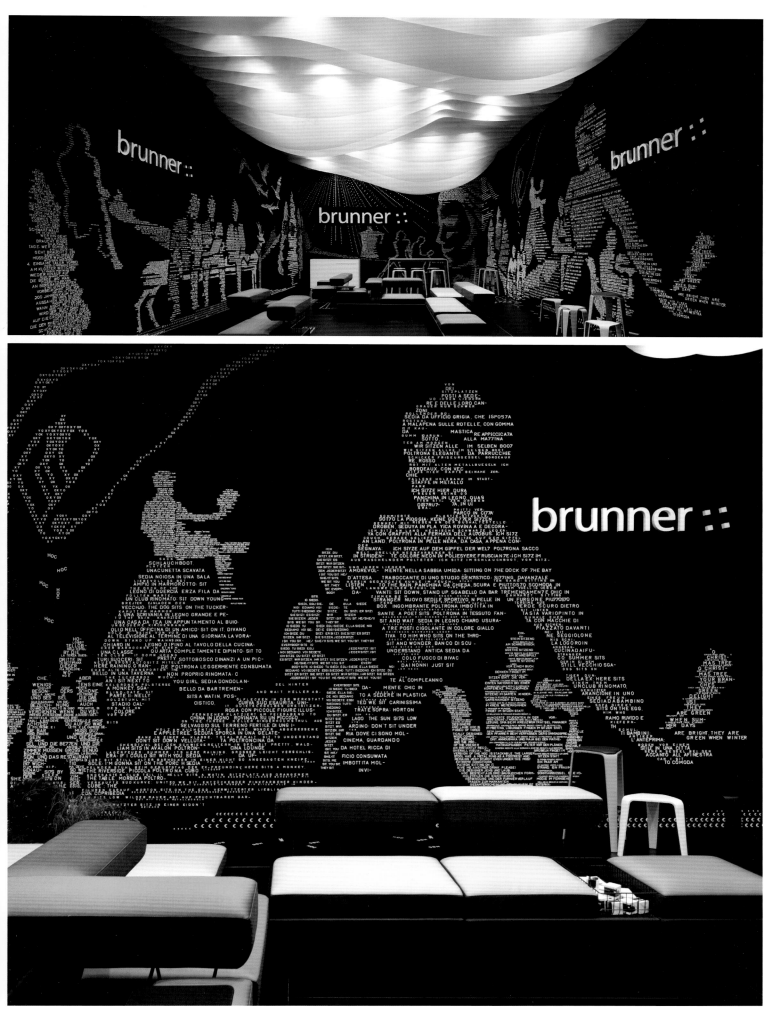

Commercial

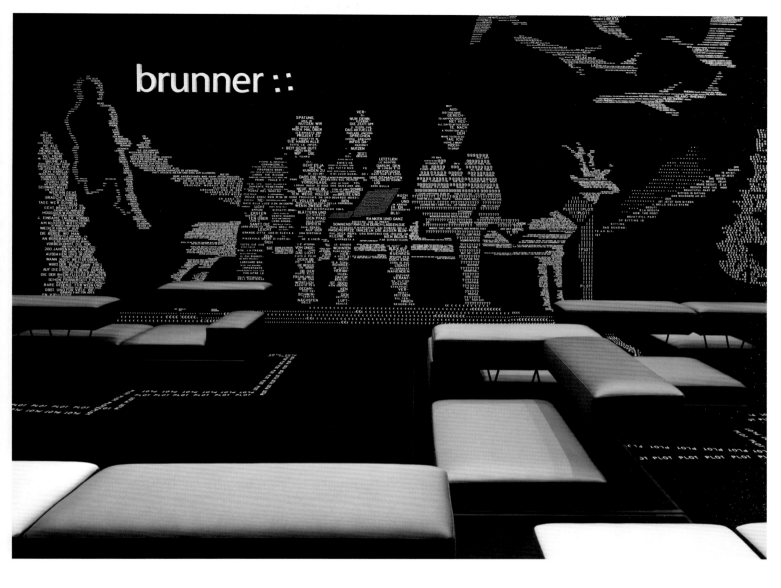

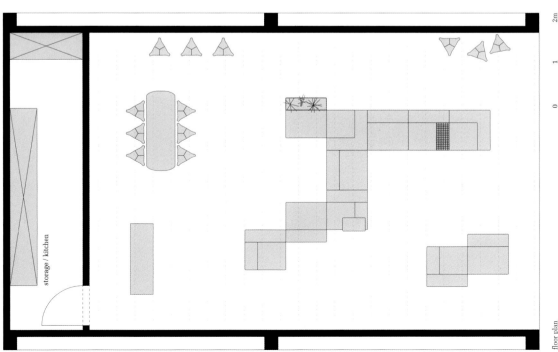

floor plan

Commercial

Janoschka Fair Stand Drupe

Design agency: Ippolito Fleitz Group GmbH
Designer: Peter Ippolito, Gunter Fleitz, Axel Knapp, Daniela Schröder, Anna Maier
Photographer: Zooey Braun
Client: Janoschka Holding GmbH
Country: Germany

Janoschka presents its international partners and potential new customers with a comprehensive spectrum of services and production competencies in the field of printing plates and reproduction. Special exhibits and communication areas convey the often quite abstract processes and services to the visitor in a fun and easy-to-understand way. The stand design features a conscious play on extreme leaps in scale, using visuals and communication derived from the world of printing plate production and reproduction. The result is a highly distinguishable overall impression for the trade fair visitor.

Janoschka 展位

Janoschka 公司向他们的国际合伙人和潜在新客户呈现公司在印刷板和复制领域的全方位服务范围和产品竞争力。特殊的展览和传播区域以好玩易懂的方式向参观者展示了抽象的流程和服务。展位设计以极端的尺寸变化为特色，使用了来自于印刷板生产和复制世界的视觉图形和传播元素。最终的设计极具辨识度，为参观者留下了深刻的印象。

设计机构：Ippolito Fleitz 设计集团　设计师：彼得·依普利托、甘特·福莱兹、阿克塞尔·克纳普、丹妮拉·施罗德、安娜·梅尔　摄影：佐伊·布朗　委托方：Janoschka 公司　国家：德国

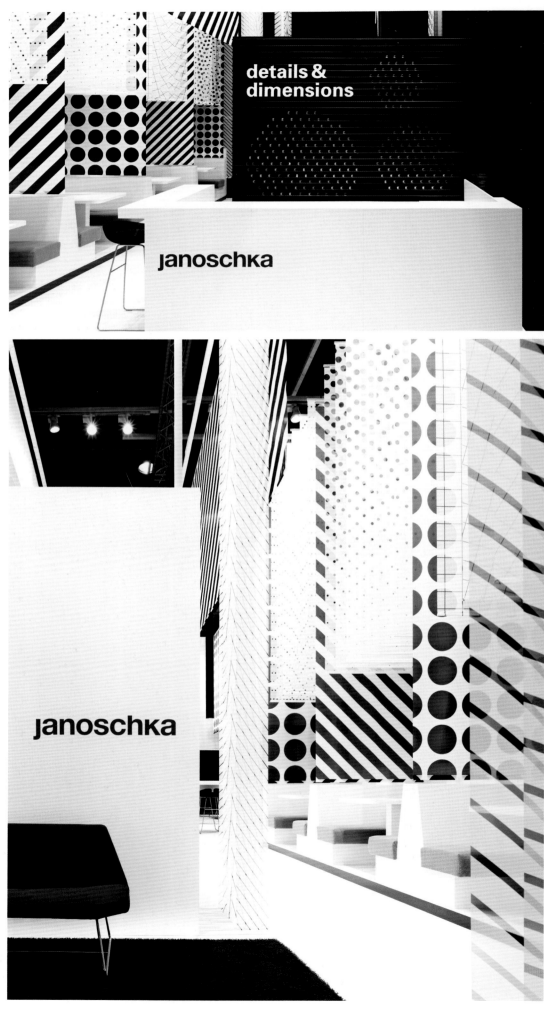

Commercial

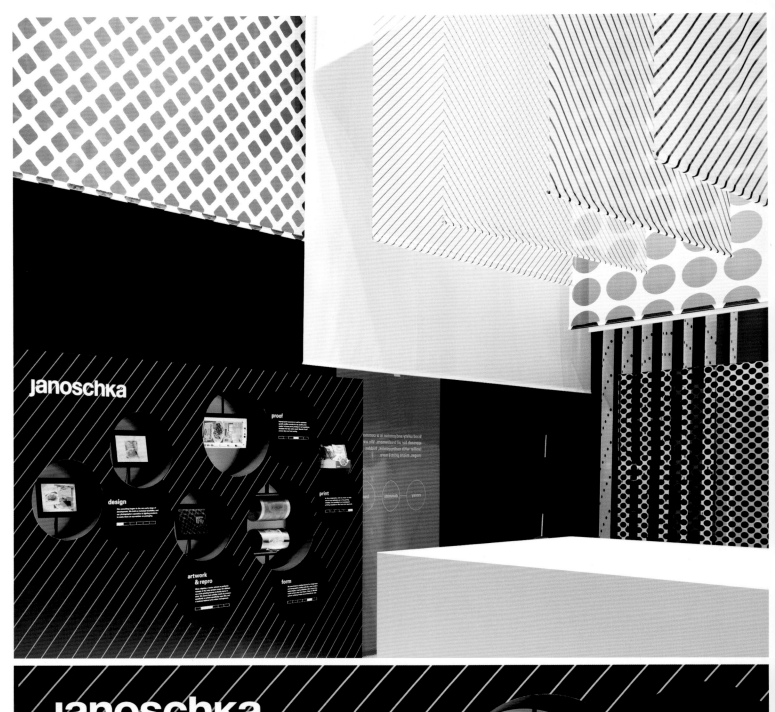

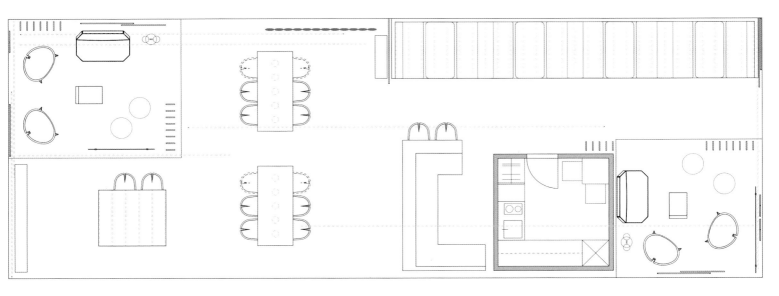

Commercial

Armstrong Fair Stand BAU 2013

Design agency: Ippolito Fleitz Group GmbH
Designer: Peter Ippolito, Gunter Fleitz, Tim Lessmann, Tanja Ziegler, Alexander Assmann, Sungha Kim
Photographer: Armstrong/P.G.Loske, Ippolito Fleitz Group GmbH
Client: Armstrong DLW GmbH
Country: Germany

The Armstrong exhibition stand functions as a communication platform and gives visual expression to the company's field of expertise. Every surface is covered by a complex, geometric pattern consisting of different cuts of Armstrong materials. A large rear wall, concealing several support rooms, carries a striking, abstract interplay of colours and shapes that create a sense of depth and perspective. From this starting point, the spatial graphic spreads out across the entire floor, covering the reception counter and conference tables. An open communication zone of tables and counters is demarcated and contained by a folded ceiling element. The latter serves as a three-dimensional counterpart to the two-dimensional structure of the spatial graphic. A polygonal podium encloses a calmer communication zone within the stand. The Armstrong exhibition stand makes a strong and striking visual impact that can be transported well. Using an installation that encompasses the space in a collage of materials, Armstrong recommends the construction material linoleum for use in contemporary, cutting-edge interior design.

Armstrong 公司展位——2013BAU 展览

Armstrong 公司的展位起到了交流平台的作用，从视觉上展示了公司的专业领域。每个空间平面都覆盖着由 Armstrong 所生产的材料构成的几何图案。一面大型背景墙将几个辅助房间隐藏起来，通过抽象的色彩与形状互动营造出一种层次感。以此为出发点，空间图形延展至整个台面，覆盖了接待台和会议桌。由桌子展台所构成的开放交流区被一个折叠天花板装置特别划分出来。三维的天花板装置与二维的空间图形结构形成了对应。一根多边形立柱将更安静的交流区包围起来。Armstrong 公司的展位营造出强烈的视觉效果，形成了良好的视觉传播。材料的拼接让 Armstrong 公司能够向人们推荐时尚、前卫的建筑材料——新型油毡布。

设计机构：Ippolito Fleitz 设计集团 设计师：彼得·依普利托、甘特·福莱兹、蒂姆·莱斯曼、塔尼亚·齐格勒、亚历山大·阿斯曼、金松佳 摄影：Armstrong/P.G.罗斯科、Ippolito Fleitz 设计集团 委托方：Armstrong 公司 国家：德国

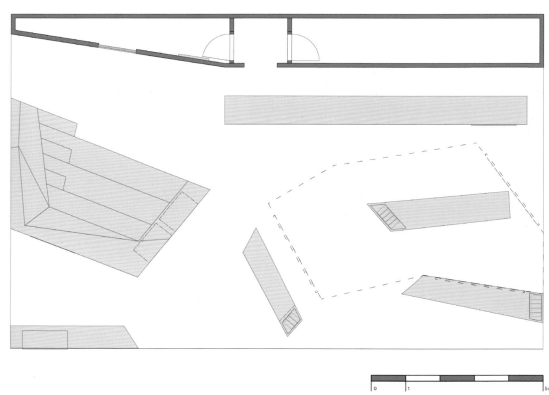

Burkhardt Leitner

Design agency: Ippolito Fleitz Group GmbH
Client: Burkhardt Leitner
Country: Germany

At leading trade fair Orgatec, Burkhardt Leitner presented the new system components of its constructiv otto range. The multiple award-winning space-within-a-space solution offers the utmost flexibility, from relatively stationary constructions for use as meeting areas and transparent yet enclosed offices for the working environment, to temporary usage at trade fairs. With three examples of these spatial solutions, the trade fair stand demonstrates the broad spectrum of possible usage.

Burkhardt Leitner 公司

Burkhardt Leitner 公司借 Orgatec 这一大型展会之机推出了它最新开发的系统建筑元件 "constructiv otto"。这一获得多项大奖肯定的 "空间中的空间" 应用极为灵活，既可以作为会议室固定在一处，也可以作为一个透明但封闭的办公空间，或者用于展会上的临时需要。公司通过这一展位设计，以三个空间案例展现出这一系统的广泛应用领域。

设计机构：Ippolito Fleitz 设计集团 委托方：Burkhardt Leitner 公司 国家：德国

Commercial

Commercial

Candy Abalone

Design agency: Liminal Graphics
Designer: Tracey Allen & Pippa Dickson
Photographer: Stuart Gibson
Client: Candy Ab
Country: Australia

The Candy Ab design was restricted by a complex site where multiple buildings were being located to a new site. The project was designed to both conceal and enhance the facilities in very simple gestures using extraordinary images of the wild Tasmanian landscape.

鲍鱼商店

鲍鱼商店的设计受到了复杂地形的限制，在新场地内聚集了多座建筑。项目设计以简单的方式隐藏并改善了整个设施的外观，在设计中融入了大量的塔斯马尼亚野外景观图片，十分震撼。

设计机构：Liminal Graphics 设计公司 设计师：特雷西·艾伦、皮帕·迪克逊 摄影：斯图加特·吉博森 委托方：鲍鱼商店 国家：澳大利亚

Commercial

NATURA HOME Showroom

Design agency: Lantavos Projects Architecture and Design
Photographer: Vice Versa
Country: Greece

Natura Home showroom was created to accommodate textile samples for furniture and curtains, exhibition space for 5 orthopedic bed mattresses, a meeting area, a repairs workshop, reception desk, and hygiene area for staff and clients. The general layout of the showroom was based on a sense of circular movement in order the staff and clients to have direct access in each segment at any time. The arcades are composed of vertical repeated panels that they do not touch on the floor giving a visual impression of floatation and that of a less enclosed space while function as separators of fabrics in categories. Each panel side has a different colour that switches from magenta to black and the opposite, depending on the direction which someone is moving in the space, creating an illusion. The section of bed mattresses appears as a separate piece of the showroom and is defined by the Greek alphabet letter 'Π' shape-form, surrounding the beds. The tones of green and brown in that section, used to emphasise a direct connection with nature, as all mattresses are made of natural materials. The vivid colours, the magenta and white resin floor, the use of mottos and words on stickers to label different sections, and the manipulation of lighting, created a brand image and entice customers to browse the space.

自然之家展示厅

自然之家展示厅汇集了家具和窗帘的布料样品、五张床垫、一个会议区、一个修补工坊、接待前台以及卫生间。展示厅的总体布局以圆圈运动为基础，让员工与客户可以随时直接到达任何部门。拱廊由反复的垂直面板构成，它们不与地面接触，从而形成了一种悬浮感，让空间不那么封闭，同时也可区分不同的面料。每块面板的侧面有两种色彩：品红和黑色，从不同的方向走来会看到不同的色彩。床垫部看起来像一个独立的展示厅，呈现为希腊字母"Π"的造型。这一部门采用了绿色和棕色来突出与自然的直接联系。鲜活的色彩、品红与白色拼接的树脂地板、短语和文字粘贴的使用以及变换的灯光效果共同营造了品牌形象，吸引着消费者巡游整个空间。

设计机构：Lantavos Projects建筑设计事务所 摄影：维斯·维萨 国家：希腊

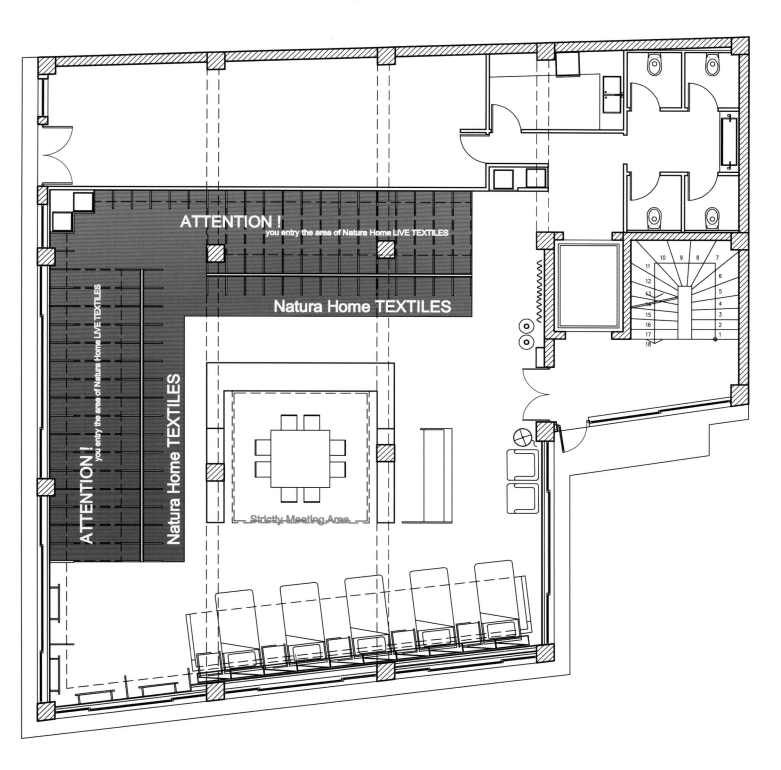

Ex.t Cersaie

Design agency: D'Apostrophe
Designer: Donatello D'Angelo, Cosimo Damiano D'Aprile
Photographer: Francesco Niccolai
Client: Ex.t
Country: Italy

D'Apostrophe goes on the development of the new identity for the young Italian brand that during Cersaie 2011 presents a new collection of young, ironic and coloured products. It's just on this features that the design of the space and the communication are developed. A series of compositions have been realised, in which the objects are "textured" on strongly coloured background.

Ex.t 展位——博洛尼亚陶瓷卫浴展

在2011博洛尼亚陶瓷卫浴展上，D'Apostrophe设计公司为这个年轻的意大利品牌打造了全新的品牌形象，他们在展览上展示了一系列年轻、多彩的卫浴产品。空间设计与传播设计全部以该品牌的产品特色为基础，从而打造了一系列出色的组合。所有产品都与彩色背景巧妙地融合起来。

设计机构：D'Apostrophe 设计公司 设计师：多纳泰罗·德安吉洛、科西莫·达米亚诺·达普利勒 摄影：弗朗西斯科·尼科莱 委托方：Ex.t 国家：意大利

Commercial

Gasworks Cellar Door

Design agency: futago_
Photographer: Peter Whyte & Futago
Client: Vantage Hotel Group
Country: Australia

The Gasworks Cellar Door, is a new business model in Tasmania, where wineries from across the state are showcased. The heritage building consists of three rooms which are dedicated to describing the particularities of each region – east, south & north – as well a central foyer describing the history of wine-making in Tasmania. The concept, developed by the architects, is for it to be a gallery-like experience where people can enjoy a glass of wine and read about Tasmanian wine. The project involved Preston Lane Architects and, Chris Viney who developed the interpretive content.

煤气厂酒窖展示厅

煤气厂酒窖展示厅是塔斯马尼亚岛上一种全新的商业模式，它展示了来自塔斯马尼亚州各地的葡萄酒。这座历史建筑中的三个房间分别展示着东、南、北三个地区的酒品，而中央大厅则用于介绍塔斯马尼亚的葡萄酒酿造史。展示厅的空间设计让人们可以尽情地参观品酒。项目的参与者还包括 Preston Lane 建筑事务所和设计说明文本的克里斯·维尼。

设计机构：futago 设计公司　摄影：皮特·怀特、Futago　委托方：Vantage 酒店集团　国家：澳大利亚

Commercial

The Apple Shed Museum and Cider House

Design agency: futago_
Photographer: Jonathan Wherrett
Client: Willie Smith Cider
Country: Australia

The Apple Shed is an ambitious project undertaken by the talented and passionate team behind Willie Smith's Organic Cider. Futago was brought onto the project by Managing Director Sam Reid and Creative Director Glen Barry, to interpret the rich history of the apple industry in the Huon Valley. The design team who worked closely with Andrew (co-Director at Willie Smith's) and Ellie Smith to bring the apple stories to life included former Redevelopment Content Manager from the Tasmanian Museum & Art Gallery Bill Seager, and writer, Chris Viney. The design was very much a collaboration with Cumulus Architects on a limited budget and short timeline to transform what was a tired, dusty and dark space into what it is today.

苹果屋博物馆和苹果酒屋

苹果屋是由威利·史密斯有机苹果酒公司发起的一项雄心勃勃的项目。公司委托 Futago 将洪恩谷丰富的苹果产业历史展示出来。设计团队与威利·史密斯公司的工作人员紧密合作，将苹果的故事展现得栩栩如生。项目设计与 Cumulus 建筑事务所共同完成，在有限的预算和期限内将一个破旧昏暗的空间打造成如今的样子，大获成功。

设计机构：futago_ 设计公司 摄影：乔纳森·威利特 委托方：威利·史密斯苹果酒公司 国家：澳大利亚

MAKING THE GRADE

AN INTERNATIONALLY-FAMOUS HERITAGE ORCHARD — IN OUR OWN BACKYARD

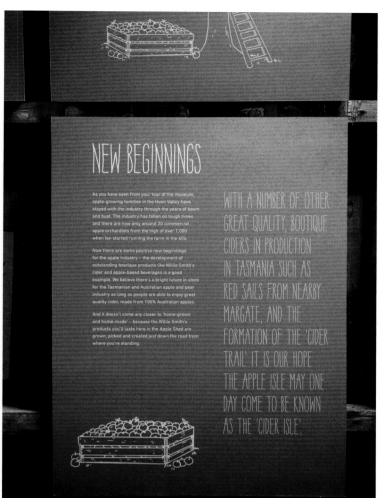

Camera di Commercio

Design agency: CCRZ + Studio Brambilla Orsoni
Designer: Eugenio Castiglioni
Photographer: Simone Cavadini
Client: Camera di Commercio Como
Country: Italy

This is a design of the interior signage system. Several buildings of diverse architectural style and eras compose the Headquarters of the Chamber of Commerce. In view of this, the elements of the signage system are designed to operate in a transversal manner throughout the building, adapting to the different functional needs but at the same time maintaining a character of neutrality and decisive.

科摩商会

本项目是一个室内导视系统设计。几座建筑年代和风格都十分迥异的建筑共同组成了商会的总部。因此，导视系统的设计元素必须以横向的方式贯穿建筑，既要适应不同的功能需求，又要保持中性、果断的设计风格。

设计机构：CCRZ + Brambilla Orsoni 工作室 设计师：欧金尼奥·卡斯蒂廖尼 摄影：西蒙·卡瓦蒂尼 委托方：科摩商会 国家：意大利

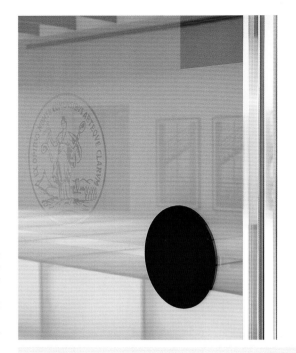

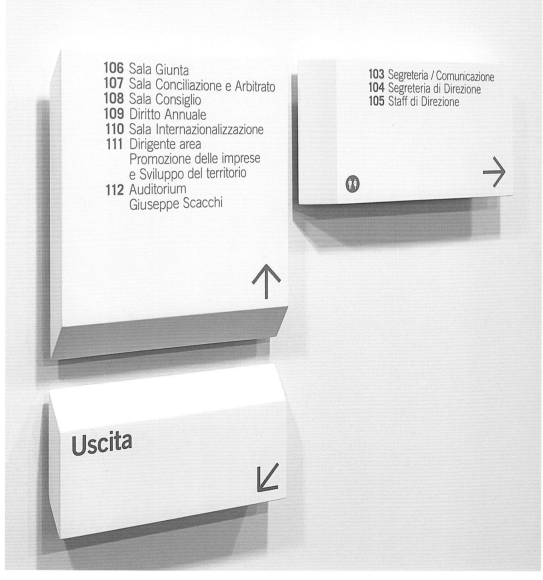

Commercial

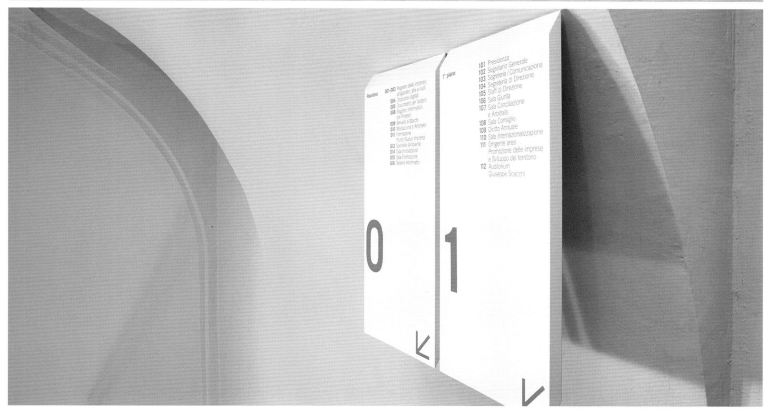

Commercial

MOME Open Day

Designer: Aron Filkey, Nora Demeczky
Photographer: Moro Mate, Daniel Racz, Gergo Gosztom, Aron Filkey, Nora Demeczky
Country: Germany

In the design of the campaign and wayfinding system of MOME's "Open Day", the primary goal has been to develop a structured, cost effective, informative and visually appealing system to aid visitors interested in applying for admission to the school. The wayfinding system also provides information about the different departmental programs and their locations. Since the campus is full with colourful decors in themselves providing strong visual stimulation, the designers have used mainly black and white negative blocks and Xerox printing. With the relatively colourful background, the black and white design appears almost subdued yet lively. In addition, its cost effectiveness has clearly appealed to the client resulting in the acceptance of the proposal. As the letter blocks come across as a grid, the designers have decided to utilise a monospace letter form choosing the Typestar font family. Finally, their artistic vision of bringing to life the purely computerised graphic design by linking it to the human is represented in the final assembly of the way-finding system reflecting the human input along with the use of human hands covered by white gloves on the banners and the website of the campaign.

MOME 学校开放日

在 MOME 学校开放日的活动及导视系统设计中，主要目标是打造一个结构化、高性价比、信息丰富且美观的系统来帮助参观者了解学校。导视系统还提供了各部门活动的信息及它们的位置所在。由于校园里布满了多彩的装饰，具有强烈的视觉刺激，设计师主要运用黑白两色和 Xerox 施乐打印来实现导航设计。在相对多彩的背景下，黑白设计看起来低调而鲜活。此外，设计的经济实用也深受客户的青睐。由于字母模块组成了一个网格，设计师决定利用单个字母的形式并选择了 Typestar 字体。最终，设计师将计算机图形变成了鲜活的导视系统，通过手工安装到横幅和活动的网站上。

设计师：阿伦·费尔齐、诺拉·德姆科斯基 摄影：莫罗·梅特、丹尼尔·拉齐、盖尔戈·古斯特姆、阿伦·费尔齐、诺拉·德姆科斯基 国家：德国

Westerdals Bachelor Exhibition

Design agency: nicklashaslestad
Designer: Nicklas Haslestad
Client: Westerdals School of Communication
Photographer: Nicklas Haslestad
Country: Norway

Westerdals graduated their first ever Bachelor students and therefore needed a full visual identity and wayfinding system. The concept was to create a celebrational visual language that striked in the cityscape, but still remained neutral in relation to the exhibited student works. The identity of the Bachelor Exhibition is inspired by Westerdals' architecture, surroundings and surfaces. Squares and boxes in various sizes with distinctive colour were placed throughout Oslo, around- and inside the building. The square format are also used in marketing materials and the signage system. Each study program had its own distinct colour which made it easy to navigate around the exhibition. The Bachelor Exhibition featured work from the department of Graphic Design, Film & TV, Texts & Copywriter, Experience & Event Design, Retail Design and Art Direction.

韦斯特达尔斯学院学士毕业设计展

韦斯特达尔斯传播学院迎来了第一批学士毕业生，因此需要全套的视觉识别和导视系统。设计理念是打造一套喜庆的视觉语言，既能在城市景观中脱颖而出，又能以中立的形式展示学生作品。学士毕业设计展的标识从学院建筑、周边环境和墙面、地面中获得了灵感。各种不同色彩、不同尺寸的方形和盒子遍布建筑内外。方形还被应用在营销材料和导视系统中。每个学习项目都有其独特的色彩，让展览导航变得更方便。学士毕业设计展的设计作品主要来自于平面设计系、电影电视系、文本文案系、体验与活动设计系、零售设计系和艺术指导系。

设计机构：nicklashaslestad 设计公司 设计师：尼克拉斯·哈斯莱斯塔德 委托方：韦斯特达尔斯传播学院 摄影：尼克拉斯·哈斯莱斯塔德 国家：挪威

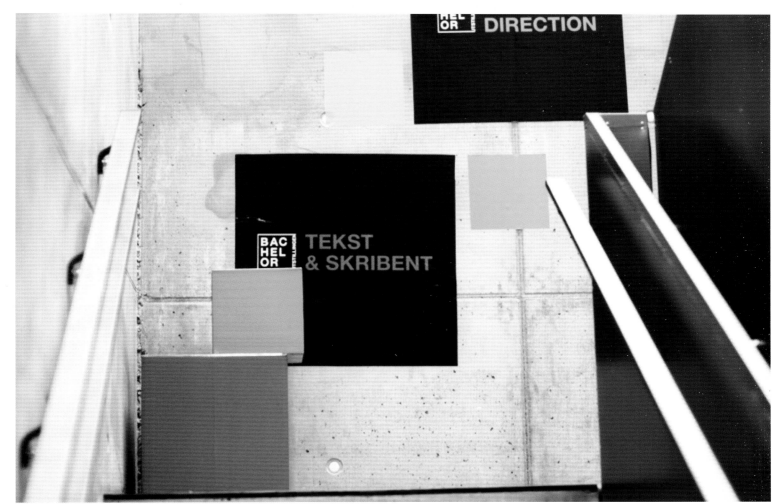

Karski. Don't Let the World Forget

Design agency: KUKI Krzysztof Iwanski
Designer: Krzysztof Iwanski
Photographer: Krzysztof Iwanski
Client: The Marek Edelman, Centre of dialogue and Tolerance in Lodz
Country: Poland

The exhibition accompanies the Jan Karski's Centenary Celebrations Birthdays in Lodz held under the honorary auspices of the President of the Republic of Poland Bronis aw Komorowski. The designer was asked to make a promotion poster for the exhibition togehter with a catalogue to it and to design all the exhibition visual concept.

卡尔斯基纪念展：别让世界忘记

本次展览与罗兹市举办的扬·卡尔斯基百年诞辰庆典共同展开，整个庆典活动受到了波兰总统布罗尼斯瓦夫·科莫洛夫斯基的大力支持。设计师受委托对展览进行了宣传海报、宣传册以及整体视觉概念的设计。

设计机构：KUKI 设计公司 设计师：克里斯多夫·伊万斯基 摄影：克里斯多夫·伊万斯基 委托方：马雷克·埃德尔曼、罗兹对话与宽容中心 国家：波兰

KARSKI.
NIE DAĆ ŚWIATU ZAPOMNIEĆ

Trzykrotnie przedzierał się z okupowanej przez Niemców Polski na Zachód, by przekazać informacje o sytuacji w kraju. Jako łącznik między polskim podziemiem a rządem polskim na uchodźstwie znajdował się w samym centrum działalności konspiracyjnej. Pojmany przez gestapo, torturowany, usiłował popełnić samobójstwo, byleby tylko nie zdradzić tajemnic konspiracji. Oddany żołnierz państwa podziemnego, ceniony za uczciwość, roztropność i odwagę. W powszechnej pamięci zapisał się jako ten, który powiedział światu o zagładzie Żydów. I jako ten, którego świat nie wysłuchał. Kim był i kim jest dla nas dzisiaj Jan Karski? Bohaterem? Świadkiem największej w dziejach zbrodni? A może – wyrzutem sumienia...

JAN KARSKI.
DON'T LET THE

Three times he ma
occupied Poland t
information on the
As a liaison betwe
Underground and
Government-in-E
centre of resistan
and tortured by t
suicide to avoid
to danger. A sol
State, highly val
prudence and c
remembered as
about the mass
one whom the
was and who is
A hero? A witne
history?
Or perhaps – a

Organizator:

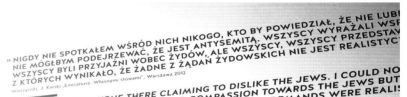

» NIGDY NIE SPOTKAŁEM WŚRÓD NICH NIKOGO, KTO BY POWIEDZIAŁ, ŻE NIE LUBI
NIE MÓGŁBYM PODEJRZEWAĆ, ŻE JEST ANTYSEMITĄ. WSZYSCY WYRAŻALI WSP
WSZYSCY BYLI PRZYJAŹNI WOBEC ŻYDÓW, ALE WSZYSCY, WSZYSCY PRZEDSTAW
Z KTÓRYCH WYNIKAŁO, ŻE ŻADNE Z ŻĄDAŃ ŻYDOWSKICH NIE JEST REALISTYC
M. Wierzyński, J. Karski „Emisariusz. Własnymi słowami", Warszawa 2012

» I DID NOT MEET ANYONE THERE CLAIMING TO DISLIKE THE JEWS. I COULD NO
OF ANTI-SEMITISM. THEY EXPRESSED COMPASSION TOWARDS THE JEWS BUT
PRESENTED ARGUMENTS WHY NONE OF THE JEWISH DEMANDS WERE REALIS
M. Wierzyński, J. Karski "Emisariusz. Własnymi słowami", Warsaw 2012

» MOŻNA TO RÓŻNIE WYJAŚNIAĆ: JEDNI SIĘ BALI, INNI DBALI O SWOJĄ KAR
OBAWIALI SIĘ, ŻE JAK ZACZNĄ ZBYT GŁOŚNO MÓWIĆ O TRAGEDII ŻYDOW
ANTYSEMITYZM. SAM NIE WIEM. SYTUACJA NIE BYŁA JASNA, PEWNE JES
ZJEDNOCZONE NIE ZROBIŁY NIC I ANGLIA TAKŻE NIE ZROBIŁA NIC. ALBO
I TO WIEMY. «
Wierzyński, J. Karski „Emisariusz. Własnymi słowami", Warszawa 2012

» THERE ARE DIFFERENT EXPLANATIONS: SOME WERE AFRAID, OTHERS DI
THEIR CAREERS, YET OTHERS BELIEVED THAT IF THEY HAD STARTED TAL
JEWISH TRAGEDY ANTI-SEMITISM WOULD HAVE INCREASED. I REALLY D
SITUATION WAS FAR FROM CLEAR, THE ONLY THING KNOWN FOR CERTA
STATES DID NOTHING AND ENGLAND DID NOTHING, EITHER. OR NEARLY
WE KNOW. «
M. Wierzyński, J. Karski "Emisariusz. Własnymi słowami", Warsaw 2012

2014 Taiwan International Documentary Festival

Design agency: UNION ATELIER
Designer: Minhan Lin, Weiche Wu & Jixian Ho
Photographer: Minhan Lin
Client: Taiwan International Documentary Festival
Country: China

The design concept of 2014 TIDF is "re-encounter reality". It means that through the Festival, we would like to create an opportunity for us to understand the past so we can bid a farewell to it, and then we will develop a new vision before re-countering "reality", which has always been the core value of documentary. The designers were commissioned to provide an overall branding and space design for the festival, to promote its image and international prestige. The designers take "circle" as a basic pattern and use yellow and black as main colours. The striking colour palette makes the event outstanding.

2014台湾国际纪录片节

2014台湾国际纪录片节的设计概念是"与现实重逢"。这意味着通过这个活动，我们将有机会回顾过去，对它说再见，然后在重逢"现实"之前创建全新的视野，这也是纪录片的核心价值。设计师受委托为本次纪录片节进行全方位的品牌及空间设计，以提升整个记录片节的形象和国际地位。设计师以"圆形"作为该项目的基本图形，选择黄色和黑色作为主色调。醒目的色彩搭配让本次活动脱颖而出。

设计机构：合聿设计　设计师：林敏汉、巫玮哲、霍集贤　摄影：林敏汉　委托方：台湾国际纪录片节
国家：中国

Raconter la Guerre (War Stories)

Designer: Margot Lombaert
Photographer: Marie Ringenbach
Client: Bibliothèque de Mulhouse
Country: France

"Tout Mulhouse Lit" is a major cultural event in the city of Mulhouse. Each year 2,500 visitors come to the library to meet their favourite authors around discussions, talks... The 2014 theme was "War in the XX and XXI Century". "War stories" is an opportunity for the public to engage with History and learn about some little known conflicts. The concept included work with a carpenter to build a sustainable system of display which could be reusable for future events. The modular panels evolve as long as the event goes; displaying each day a different programme, orientation according to the talks and quotes from the invited authors. The wide screens provide a more active look and feel. The public can for instance interact by hanging messages related to peace. The wide screens also display strong imagery of soldiers and civilians as well as a timeline presenting statistics about the most devastating wars since 1900.

战争故事

"米卢斯的一切"是米卢斯市的重要文化活动。每年有2,500名游客到图书馆来与他们所喜爱的作者会面，进行讨论和会谈。2014年的主题是"20与21世纪的战争"。"战争故事"让公众有机会参与到历史中，了解一些不为人知的冲突。设计包含一套木工制品，打造了可持续的展示系统，未来可以回收使用。模块化展板可以随着活动变化，每天展示一个不同的主题，随着特邀作者的会谈做出改变。宽大的屏幕给人以活跃的观感。公众可以通过悬挂留言进行互动。宽大的屏幕上还展示了士兵与平民的照片，通过时间线展示自1990年以来最具毁灭性的战争。

设计师：玛格特·罗姆贝尔特 摄影：玛丽·林根巴赫 委托方：米卢斯图书馆 国家：法国

Culture

The 125th Assembly of the Inter-Parliamentary Union

Design agency: Intégral Ruedi Baur
Art direction: Ruedi Baur, Axel Steinberger
Designer: Denis Coueignoux, Janine Hofmann, Lisa Jacob, Angelina Köpplin
Client: Swiss confederation Parliament Services, Berne
Country: Switzerland

The fair's location in Berne is a large-scale facility faced with an obvious lack of atmospheric qualities and amenity values. Responding to theses deficits, Integral interpreted the task of a temporary signage system for the 125th IPU Assembly as a spatial and scenographic design that welcomes and accompanies the assembly guests. Within three months, Integral developed and realised a visual language for the complete event communication. The created design system with essential elements – typography, cartography and information graphics – was fully applicable to the needs of 2D- as well as 3D-media. In terms of scenography, information graphics became the main tool to reorganise and rethink the facilities. Integral developed a graphical system with the ability to individually adapt to dimensions and media, that was based on IPU facts and data showing and comparing main aspects such as members, population, economic and democratic forces. Consequently, the classic signage system was replaced by a family of various applications rescaling and transforming the facilities into lounges and gallery spaces to nurture the discourse and to represent the event's glance to the outside world.

第 125 届议会联盟大会

会展的所在地是一座大规模设施，明显缺乏背景气氛和舒适感。为了应对这些缺陷，设计师将第 125 届议会联盟大会的导视设计堪称是一个空间及场景设计，为与会人员提供了友好的氛围。在 3 个月内，设计师开发并实现了整个会场的视觉识别系统。以字体、地图和信息图形等基本元素构成的设计系统完全适用于二维及三维媒体。在场景设计中，信息图形成了改造设施的主要工具。设计师开发了一套可独立使用的图形系统，该系统根据议会联盟的数据显示并对比了成员、人口、经济和民主力量等主要方面为基础。最终，经典的导视系统被各种不同尺度的导视图形所取代，成功改造了各种空间，体现了活动的国际性视野。

设计机构：Intégral Ruedi Baur 设计公司 艺术总监：吕迪·鲍尔、阿克塞尔·施泰因贝格尔 设计师：丹尼斯·库伊格诺、亚尼内·霍夫曼、丽萨·雅各布、安吉丽娜·库柏林 委托方：瑞士联邦议会服务部 国家：瑞士

Culture

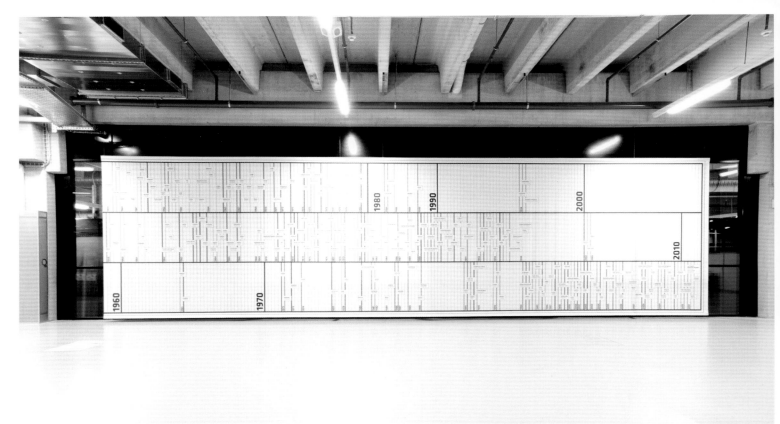

Culture

Advancing The Mission: The Cooper Union At 150

Design agency: Lee H. Skolnick Architecture + Design Partnership
Client: Advancing the Mission
Country: USA

As part of its 150th anniversary and overall institutional rebranding, the Cooper Union retained LHSA+DP to design Advancing the Mission, a new permanent exhibition in the lobby of its influential Great Hall. Widening the space's volume, and honoring the Great Hall's historically significant architecture, the exhibit invites visitors to discover the Hall's role in promoting the institution's growth through ideals of democracy, social justice, philanthropy, and public education. An encompassing red drum creates a meditative entryway, taking its shape from the building's enigmatic elevator, and quotes from the Union's founder wrap pillars supporting the architecture: "We are all bound up in one common destiny." Glass panels double as layered graphic surfaces and protective casework for artifacts, and wall paper graphics set a relevant backdrop for presenting the stories of iconic radicals who took the Great Hall stage over the years, including Abraham Lincoln, Susan B. Anthony, and Frederick Douglass. Visitors to the exhibit are encouraged to recognise that many cultural shifts of national scale were launched from this very site through democratic community activism.

"改良任务"展览：库伯联盟学院 150 周年庆典

作为库伯联盟 150 周年庆典和整体品牌重塑的一部分，库伯联盟委托 LHSA+DP 设计在学院大会堂举办了"改良任务"展览。展览拓展了空间，纪念了大会堂的历史意义，邀请参观者共同发现大会堂在促进学院发展中所扮演的角色。鲜红的大鼓营造出令人沉思的入口通道，它的造型来自于建筑神秘的电梯，引用了联盟的创始人的话："我们拥有共同的命运。"玻璃板既是分层的图形展示平面，又能保护内部的收藏品。墙纸图形为展示提供了背景，先后呈现了亚伯拉罕·林肯、苏珊·B·安东尼和弗雷德里克·道格拉斯等标志性活动家的生平。展览让参观者了解到许多国家层面上的文化变迁都是从这里通过民主社群活动开始的。

设计机构：LHSA+DP 设计公司 委托方："改良任务"展览 国家：美国

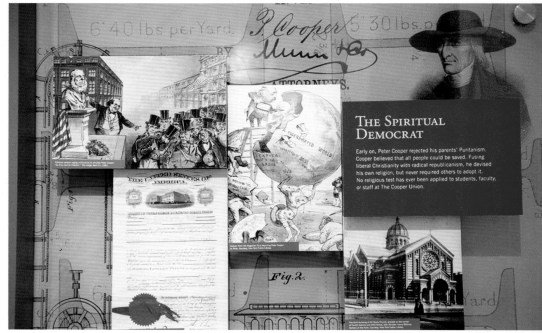

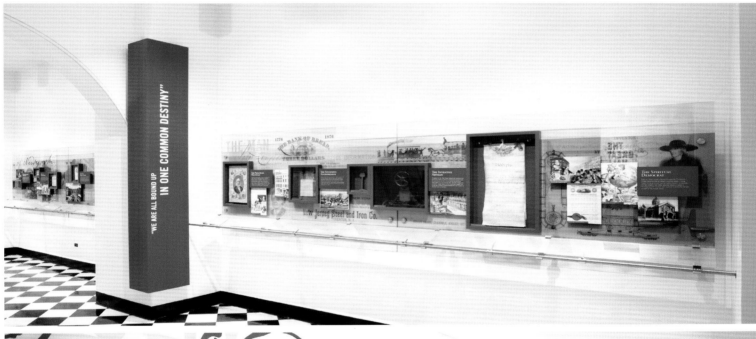

Culture

Culture City of East Asia 2014 YOKOHAMA

Design agency: NOSIGNER
Art director: Eisuke Tachikawa
Client: Yokohama Creative City Centre (Ycc)
Country: Japan

Culture City of East Asia 2014 YOKOHAMA is a program for cultural exchange between Japan, China and Korea. Based on the theme "one-day airport", NOSIGNER designed an interactive space by using lights and projections that react to sound. The info-graphics of the three cities inspire people to think of cultural exchange.

横滨 2014 东亚文化城市展

2014 横滨东亚文化城市展是一个旨在促进日本、中国和韩国文化交流的项目。设计师以"一日机场"为主题，利用灯光和声控投影设计了一个互动空间。三个城市的信息图形激发人们对文化交流进行思考。

设计机构：NOSIGNER 设计公司　艺术总监：太刀川英辅　委托方：横滨创意城市中心　国家：日本

Culture

Estand Festival de la Infància

Design agency: Petit Comitè
Art director: Yolanda Martin
Photographer: Quim Massana
Client: Government of Catalonia
Country: Spain

The project was done for a stand in the Youth Festival from Barcelona (Salón de la infancia de Barcelona). The illustrations and typography are "mequetrefe" a font designed in collaboration with the illustrator Juanjo Saez and the designer Narcis Sauleda. Petit Comitè started with Juanjo's basic calligraphy and soon they saw the need and the wish to make it grow into a more complete system, with bold versions, outline, volume, shadows, etc. and of course some icons with the theme: all you may need to find your way round in a city.

儿童节展位

项目是为巴塞罗那儿童节所设计的展位。展位设计所使用的插画和文字选择了插画师胡安乔·赛斯和设计师纳西瑟斯·索勒达共同设计的mequetrefe字体和图形。Petit Comitè 以胡安乔的基本书法为出发点，随后将其打造为一个更完整的系统，加入了大胆的改造、勾勒、造型和阴影，同时也加入了主题图标：在城市中寻路的必要元素。

设计机构：Petit Comitè 设计公司 艺术总监：尤兰达·马丁 摄影：奎姆·马萨纳 国家：西班牙

MAS Jane Jacobs & the Future of New York

Design agency: karlssonwilker inc.
Client: The Municipal Art Society
Country: USA

The Municipal Art Society's Jane Jacobs and the Future of New York exhibition explored the legacy of the renowned urban activist and the urban design principles presented in her classic text, The Death and Life of Great American Cities. karlssonwilker designed the exhibition, the accompanying website, and outdoor print campaign.

简·雅各布斯 & 纽约未来展

城市艺术协会所举办的简·雅各布斯 & 纽约未来展探索了这位著名城市活动家为我们留下的遗产以及她的经典文本《美国城市的生与死》中的城市设计法则。karlssonwilker 设计公司对展览、对应的网站以及户外印刷宣传进行了整体设计。

设计机构：karlssonwilker 设计公司 委托方：城市艺术协会 国家：美国

Friedliche Revolution

Designer: Christian Pannicke
Photographer: Christian Pannicke
Client: Stasi Museum Berlin
Country: Germany

This project is the exhibition and wayfinding design for the exhibiton "Friedliche Revolution" at the forecourt of the Stasi Museum. The exhibition take place for the 25-year anniversary of the fall of the Berlin Wall. Who does not move, doesn't feel the fetters." This quote from Rosa Luxembourg is the guideline of the design concept, which combines the essence of the exhibition to the Peaceful Revolution with the Stasi museum in the face of history. Between 1961 and 1989, the resistance against the system became stronger – the Berlin Wall lost its stability. The designer uses this fact also in the design. He breaks up words and thus create a seemingly resolution language with a revolutionary character. The first massive wall finally loses itself in an unstable situation. To draw attention to the issue, existing conditions are already being used and transformed. It can be used lantern pillars, poles, fences and billboards. The existing information boards get a new colour scheme and become an orientation and information point for the visitors.

弗雷德里希革命展

项目是为在斯塔西博物馆举办的"弗雷德里希革命"展览所进行的展览及导视设计。展览是为了庆祝推倒柏林墙25周年而举办的。设计师以罗莎·卢森伯格的名言"不迁移的人感受不到束缚"为指导理念，将展览的本质与和平革命和历史结合起来。在1961年至1989年之间，人们对柏林墙的反对之声日渐强烈。设计师将这一事实融入了设计之中。他将单词断开，创造出一种看似革命性的新语言。为了体现柏林墙即将倒塌的情景，设计师运用了灯柱、细杆、栅栏、宣传牌等多种元素。原有的信息板被换上了新的色彩主题，为参观者提供了有效的导航和信息指示。

设计师：克里斯蒂安·帕尼克 摄影：克里斯蒂安·帕尼克 委托方：柏林斯塔西博物馆 国家：德国

Culture

Presseum

Design agency: Practice and Workroom
Photographer: Yong-Kwan Kim, Practice
Client: The Dong-A Ilbo
Country: Korea

Practice and Workroom developed the identity system, exhibition graphics, signage, and print applications for Presseum, the museum of Korean newspaper and its histories.

新闻博物馆

Practice and Workroom 设计公司为新闻博物馆进行了视觉识别系统、展览图形设计、导视系统以及印刷宣传品设计。新闻博物馆是专门展示韩国新闻报纸及其历史的博物馆。

设计机构：Practice and Workroom 设计公司 摄影：金永关、Practice 摄影公司 委托方：东亚日报 国家：韩国

Culture

Culture

0618 Exhibition

Design agency: CCRZ + Studio Brambilla Orsoni
Designer: Paolo Brambilla, Marco Zürcher
Photographer: CCRZ
Client: Archivio Cattaneo
Country: Italy

The exhibition in the ancient Villa del Grumello in Como recalls one of the greatest expressions of Larian rationalism. The itinerary illustrates the reclamation and requalification of the building planned by Cesare Cattaneo (ULI) adjacent to Terragni's Casa del Fascio. The title "0618" derives from the numerical ratio of the golden section, a principle implemented by the rationalists. The exhibition design is exclusively realised with double cardboard panels, which define spaces and paths within the large halls of the Villa. The studio executed the graphic design for the exhibition and catalogue; the idea and concept was developed by architects Paolo Brambilla, Renato Conti and Corrado Tagliabue.

0618 展览

这个科摩一座古宅里举办的展览展示了这座由切萨雷·卡塔内奥所设计的建筑的历史。标题 "0618" 来自于黄金分割点数值比例，黄金法则是理性主义者最常用的法则之一。展览设计全部采用双层硬纸板，利用纸板在住宅大厅里塑造空间和路线。CCRZ 担任了展览和参展名录的图形设计；而基本设计概念则由建筑师保罗·布兰比拉、雷纳托·康迪和科拉多·塔利亚布埃开发。

设计机构：CCRZ + Brambilla Orsoni 工作室 设计师：保罗·布兰比拉、马尔科·苏赫 摄影：CCRZ 委托方：卡塔内奥档案馆 国家：意大利

Culture

La Suda – El Castell del Rei

Design agency: Estudi Conrad Torras & La Petita Dimensió
Designer: Conrad Torras, Gerard Gris & Eva Pérez
Photographer: Conrad Torras
Client: Turisme de Lleida
Country: Spain

Often ignored, not only by the general public but also by the citizens of Lleida, this permanent exhibition aims to raise awareness of this castle as an important place in the evolution of the different civilizations that have inhabited this territory as well as an important enclave of Catalan history. In the only room that is left of the old castle, the visitor is invited to know the history of this place. The tour is organised through three large hanging modules, inspired by medieval lamps that host the contents inside. Using a simple pulley system, the lamps can be raised, leaving free room for holding any event, one of the project requirements. On the roof, a stunning viewpoint offers privileged views across the region. Some viewers allow the visitors to compare the present and the past of this landscape through archival photographs placed inside. The exhibition is permanent but the entire assembly is designed as an easily renewable content.

苏达国王城堡

不仅是普通民众，甚至莱里达的市民都经常忽视这个城堡。这个永久性展览旨在提升城堡的辨识度，展现它在不同文明演化中所扮演的重要角色，体现它在加泰罗尼亚历史中的重要价值。在唯一一间未经改动的旧城堡房间，游客将了解这里的历史。整个参观旅程通过三个大型悬挂模块展开，它们的设计灵感来自于中世纪内部含有信息的灯笼。简单的滑轮系统让灯笼可以上升，空出房间举办其他活动。从屋顶上可以俯瞰整个区域的美景。一些展示让游客可以将现在与过去的风景进行对比。展览是永久性的，但是整个配套设计都可以进行方便的更新。

设计机构：Estudi Conrad Torras 设计公司、La Petita Dimensió 设计公司　设计师：康拉德·托拉斯、杰拉德·格里斯、伊娃·佩雷斯　摄影：康拉德·托拉斯　委托方：莱里达旅游局　国家：西班牙

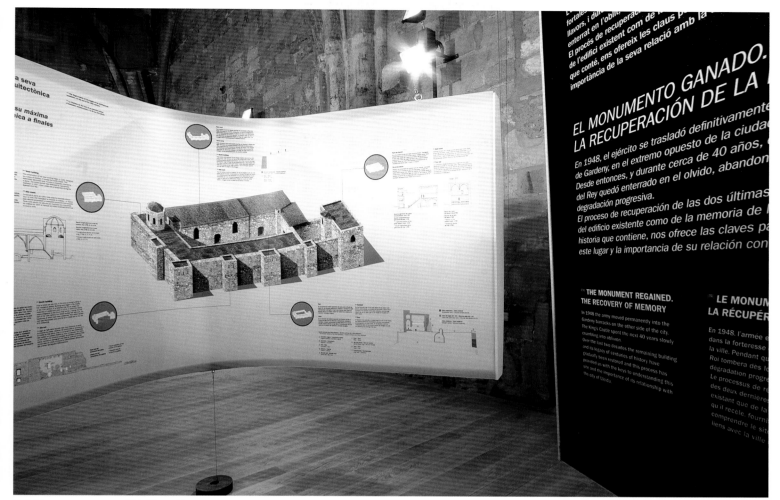

Culture

Culture

duPont Testamentary Trust Gallery

Design agency: Gresham, Smith and Partners
Designer: Jim Alderman and Tim Rucker
Photographer: Sue Root Photography
Client: duPont Testamentary Trust
Country: USA

The Alfred I. duPont Testamentary Trust, based in Jacksonville, Florida, was established by the estate of Alfred I. duPont. Through the Nemours Foundation, the trust supports a network of children's hospitals, clinics, and research facilities. Gresham, Smith and Partners provided interior design, environmental graphic design and exhibit design services for the new Trust headquarters in downtown Jacksonville. Located on the river, the building includes a dramatic four-storey atrium and a gallery space overlooking the water. The GS&P environmental graphics group worked with the members of the board of trust to create a legacy space celebrating the lives of Alfred I. duPont, his wife Jessie Ball duPont, and her brother Ed Ball. The gallery includes large illuminated glass portraits, video displays, artifact cases, and a mural reproduction of the first Nemours Children's Institute.

杜邦遗嘱信托展示厅

总部设在美国佛罗里达州的杜邦遗嘱信托由阿尔佛雷德·杜邦建立。通过内穆尔基金，杜邦信托赞助支持了一系列的儿童医院、诊所和科研机构。GS&P设计公司为杜邦信托位于杰克逊维尔市中心的新总部提供了室内设计、环境图形设计和展览设计服务。这座建在河上的建筑拥有一个四层高的中庭和一个俯瞰水面的展示空间。GS&P 的环境图形团队与杜邦信托的董事会成员共同为阿尔佛雷德·杜邦以及他的夫人杰西·鲍尔·杜邦、妻弟艾德·鲍尔打造了一个介绍他们生平的展示空间。展示厅内设有大幅发光玻璃肖像、视频展示、遗物陈列柜以及一幅首家内穆尔儿童机构的复制壁画。

设计机构：GS&P 设计公司 设计师：吉姆·奥尔德曼、蒂姆·洛克 摄影：Sue Root 摄影 委托方：杜邦遗嘱信托 国家：美国

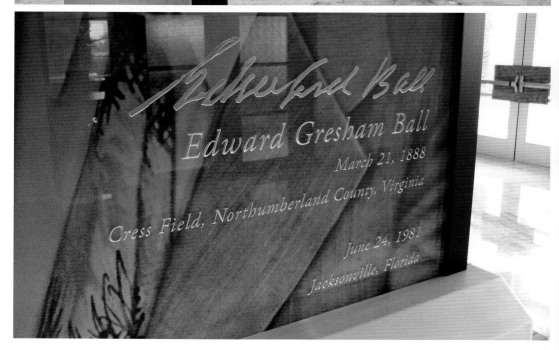

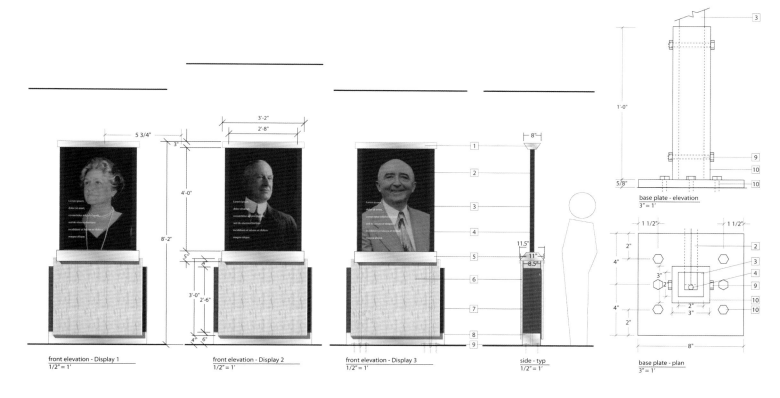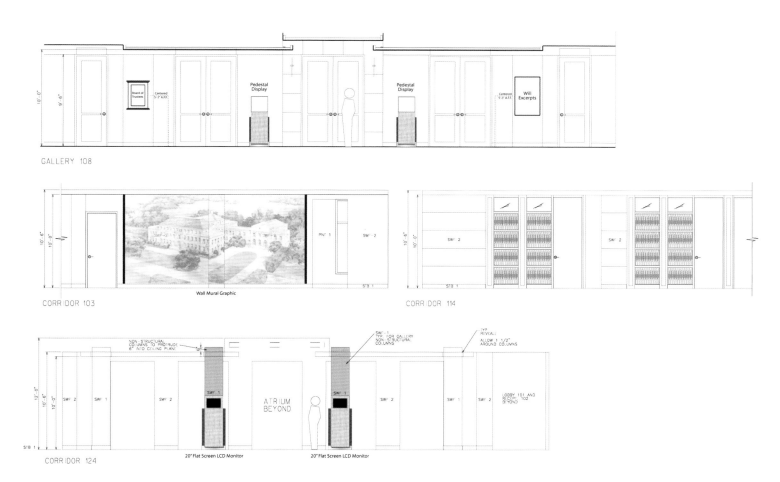

Urban Fabric

Design agency: Sasaki Associates, Inc.
Client: Sasaki Associates, Inc.
Country: USA

Urban Fabric is an interdisciplinary research project that investigates the social, environmental, and economic context of the post-industrial American city. In response to the challenges discovered in the research cities, Urban Fabric presents a selection of case studies from around the world that can serve as strategies for positive change in post-industrial American cities. On a variety of scales – from the multinational to the site-specific – these strategies are examples of successful policy, planning, and design initiatives that transform the challenges of industrial urbanism into sustainable trajectories for the future. The six-week lecture series and exhibition, hosted by the research team at the Sasaki office, began in April of 2011. The event brought together the team's independent analysis with commentary from academics, practitioners, and policy experts from across the country, attracting hundreds of participants interested in the topic of the post-industrial American city.

城市脉络展

"城市脉络"是一个跨学科研究项目,它调查了后工业化美国城市的社会、环境和经济现状。为了应对被调研城市中所发现的问题,"城市脉络"呈现了来自各地的案例研究,以求为后工业化美国城市的积极变化提供策略。从跨国策略到特定场所策略,这些规模不同的策略是成功的政策、规划和设计的典范,它们将工业化都市所面临的挑战改造成了未来的可持续发展轨迹。持续6周的讲座和展览由Sasaki设计公司的研究团队举办,于2011年4月正式开始。活动汇集了团队的独立分析,并且配有来自各国的学者、实践人员和政策专家的评论,吸引了数百名对后工业化美国城市主题感兴趣的参与者。

设计机构:Sasaki设计公司 委托方:Sasaki设计公司 国家:美国

EU 2012

Design agency: Idéskilte Horsens ApS
Client: EU 2012
Country: Denmark

EU countries take turns to chair the Council of Ministers for six months at a time. The country holding the presidency, has the task of organising and chairing its meetings. For 3 months in 2012, the city of Horsens, Forum Horsens Hotel Opus Horsens hosts for this presidency. In order to provide a good signage system for these meetings, the designers analysed the site and interviewed the users. The specially-designed signs are extensively used in the signage system. The signboard is basically white, with red, grey and black informative figures and text, legible even from a long distance.

欧盟 2012

欧盟国家每6个月轮换主持一次部长会议。主办国家的任务是组织并主持各种会议。丹麦霍森斯市在霍森斯奥普斯酒店承办了2012年历时3个月的会议。为了为本次会议打造良好的导视系统，设计师对场地特性进行了研究，并听取了使用者的意见。设计师为本次会议设计了专门的标志并在导视系统中得到广泛应用，引导标示的板材以白色为基础，图形和信息采用醒目的红色、灰色和黑色，即使在远处也清晰可见。

设计机构：Idéskilte Horsens ApS 设计公司 委托方：欧盟2012 国家：丹麦

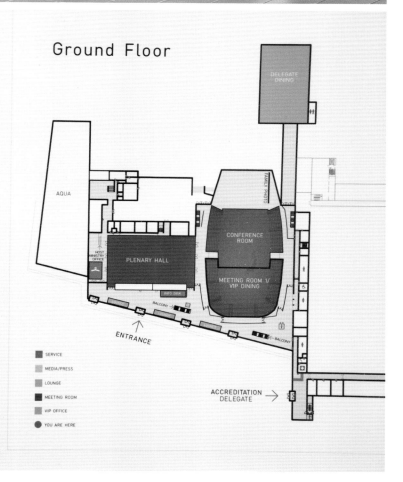

Culture

AT'12
Premio Architettura
Territorio Fiorentino

Design agency: D'Apostrophe
Designer: Donatello D'Angelo, Cosimo Damiano D'Aprile
Photographer: D'Apostrophe
Client: Fondazione Professione Architetto
Country: Italy

D'Apostrophe provides the Identity and Exhibition Design of AT'12, first edition of the prize for the best architecture on the Florentine District. The prize was divided into five categories (New Construction – Restoration – Exhibition and Interior design – Landscape Architecture – Debut) and was judged by an authoritative jury. The editorial design, as well as all the identity of the event, was inspired by the Radical Architecture that was born exactly in Florence in the Sixties of last Century.

AT'12 佛罗伦萨建筑奖展览

D'Apostrophe 设计公司为AT'12 第一届佛罗伦萨地区最佳建筑奖展览提供了形象及展览设计。奖项设置分为五个类别（新建筑、修复建筑、展览及室内设计、景观建筑、处女作），由权威评委进行评审。展览的编辑设计以及所有相关的形象识别设计灵感全部来自于20世纪60年代起源于佛罗伦萨的激进建筑风格。

设计机构：D'Apostrophe 设计公司 设计师：多纳泰罗·德安吉洛、科西莫·达米亚诺·达普利勒 摄影：D'Apostrophe 设计公司 委托方：意大利建筑职业基金会 国家：意大利

Culture

AT'14
Premio Architettura Territorio Fiorentino

Design agency: D'Apostrophe
Designer: Donatello D'Angelo, Cosimo Damiano D'Aprile
Photographer: D'Apostrophe
Client: Fondazione Architetti Firenze
Country: Italy

D'Apostrophe provides the Identity and Exhibition Design of AT'14, second edition of the prize for the best architecture on the Florentine District.

AT'14佛罗伦萨建筑奖展览

D'Apostrophe设计公司为AT'14第二届佛罗伦萨地区最佳建筑奖展览提供了形象及展览设计。

设计机构：D'Apostrophe设计公司 设计师：多纳泰罗·德安吉洛、科西莫·达米亚诺·达普利勒 摄影：D'Apostrophe设计公司 委托方：意大利建筑职业基金会 国家：意大利

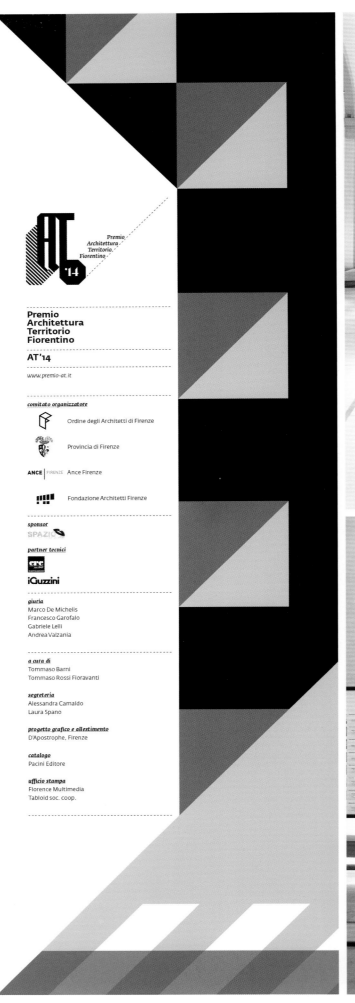

**Premio
Architettura
Territorio
Fiorentino**

AT'14

www.premio-at.it

comitato organizzatore
- Ordine degli Architetti di Firenze
- Provincia di Firenze
- ANCE Firenze
- Fondazione Architetti Firenze

sponsor
SPAZIO

partner tecnici
CRC
iGuzzini

giuria
Marco De Michelis
Francesco Garofalo
Gabriele Lelli
Andrea Valzania

a cura di
Tommaso Barni
Tommaso Rossi Fioravanti

segreteria
Alessandra Camaldo
Laura Spano

progetto grafico e allestimento
D'Apostrophe, Firenze

catalogo
Pacini Editore

ufficio stampa
Florence Multimedia
Tabloid soc. coop.

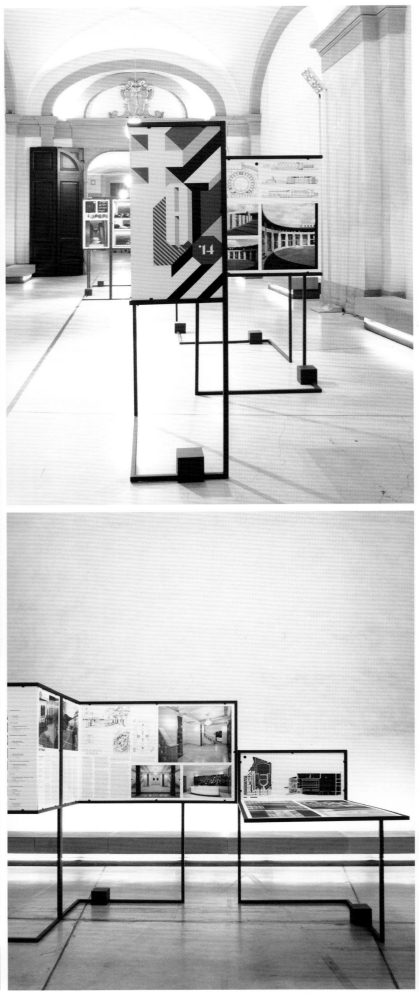

Frank Lloyd Wright and The City

Design agency: Department of Graphic Design and Advertising, MoMA
Creative director: Hsien-yin Ingrid Chou
Designer: Samuel Sherman, Tony Lee
Photographer: Martin Seck
Country: USA

The exhibition Frank Lloyd Wright and the City: Density vs. Dispersal highlights Wright's complex relationship to the city. The material reveals Wright as a compelling theorist of both its horizontal and vertical aspects. He worked simultaneously on radical new forms for the skyscraper and on a comprehensive plan for the urbanisation of the American landscape titled "Broadacre City." His work, in this way, is not only of historic importance but of remarkable relevance to current debates on urban concentration. Inspired by his explorations between vertical and horizontal aspects in his work, the designers designed a title wall to echo that relationship.

弗兰克·劳埃德·赖特与城市

"弗兰克·劳埃德·赖特与城市：密度VS分散"展览突出了赖特与城市的复杂关系。展览资料展示了赖特作为一个理论学家的卓越成就。他既能打造激进的新形式摩天大楼，又能为美国景观"广亩城市"制作综合的城市规划。他的作品不仅具有历史意义，而且与城市集中化的理论讨论相联系。设计师从他在纵向和横向空间上的探索中获得了灵感，打造了一面标题墙来呼应这些关系。

设计机构：MoMA 平面设计与广告部 创意总监：希因·英格丽德·周 设计师：塞缪尔·谢尔曼、托尼·李 摄影：马丁·塞克 国家：美国

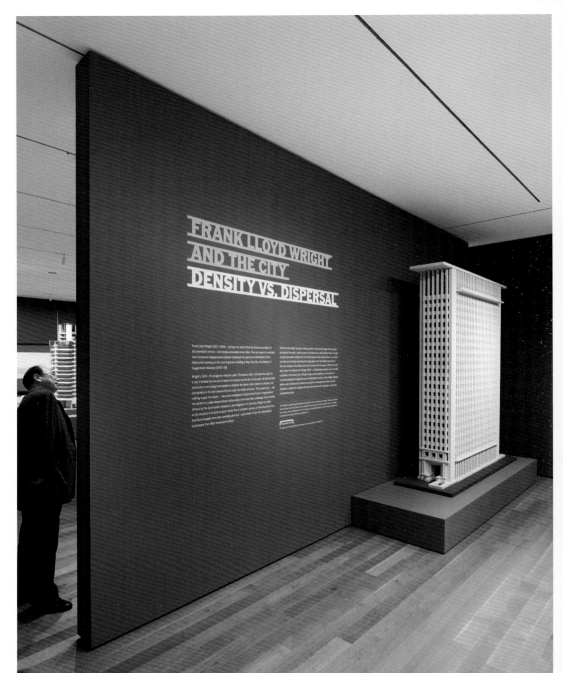

Culture

Sisters of St Joseph

Design agency: futago_
Photographer: Jonathan Wherrett
Client: Sisters of St Joseph
Country: Australia

futago_ worked with architect Greg Luttrell to develop a series of exhibitions which tell the story of the Sisters of St Joseph in Tasmania. A range of media was used including; screenprinting, video & audio (by Solid Orange) and objects in unique display cases.

圣约瑟夫修女

futago_ 设计公司与建筑师格雷格·勒特雷尔联合打造了一系列介绍塔斯马尼亚岛圣约瑟夫修女的展览。展览使用了各种媒介，包括丝网印刷、视频与音频（Solid Orange 提供）以及陈列在独特展柜中的物品。

设计机构：futago 设计公司 摄影：乔纳森·威利特 委托方：圣约瑟夫修女会 国家：澳大利亚

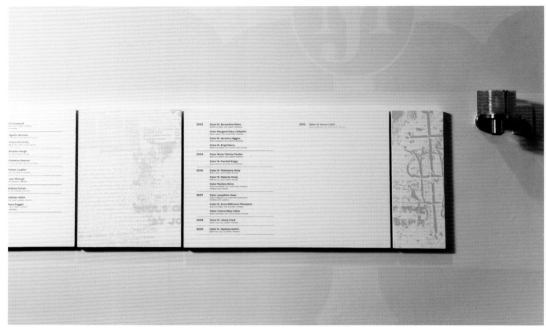

A life of simplicity lived in community...

Culture

A life of simplicity lived in community...

The Rule

Culture

The Grad Show

Design agency: Student Committee of The Grad Show
Designer: Sim Kok Huoy, Terence Yeung, Ho Hui May, Alvin Koh, Zann Lim, Steven Koswara Dahlia Loren, Christopher AP, Elsie Chong, Jiahui Lee, Keshia Anindita, Audi Irwantoro, Alim Chandra
Photographer: Steven Koswara, Zann Lim
Client: The Grad Show, Nanyang Academy of Fine Arts
Country: Singapore

The Gradshow is the name of the graduation show of Nanyang Academy of Fine Arts. Showcasing students' work at the end of their studies from BA (Hons) Graphic Communication and Diploma in Design&Media. A small team of graduating students was tasked with the whole branding process. Fun, energetic, colourful and adaptable are some of the keywords that inspired this identity. Once the basic brand identity is defined, the versatile secondary graphics were then morphed into wayfinding system in this entire exhibition.

毕业展

本项目为新加坡南洋美术学院的毕业展，展出了图形传播和设计媒体专业毕业生的毕业作品。一个由毕业生组成的团队担当了整个品牌营销流程的设计工作。有趣、有力、多彩、灵活是整个设计的关键词。在确定了基本品牌识别设计之后，通用的次级图形就被转换到了整个展览的导视系统中。

设计机构：毕业展学生会 设计师：辛戈惠、特伦斯·杨、霍慧梅、艾文·柯、查恩·林、史蒂文·科斯瓦拉、达利亚·劳伦、克里斯多夫·AP、艾尔希·庄、李佳慧、柯西亚·安妮蒂塔、奥迪·伊万托罗、阿利姆·钱德拉 摄影：史蒂文·科斯瓦拉、查恩·林 委托方：南洋美术学院毕业展 国家：新加坡

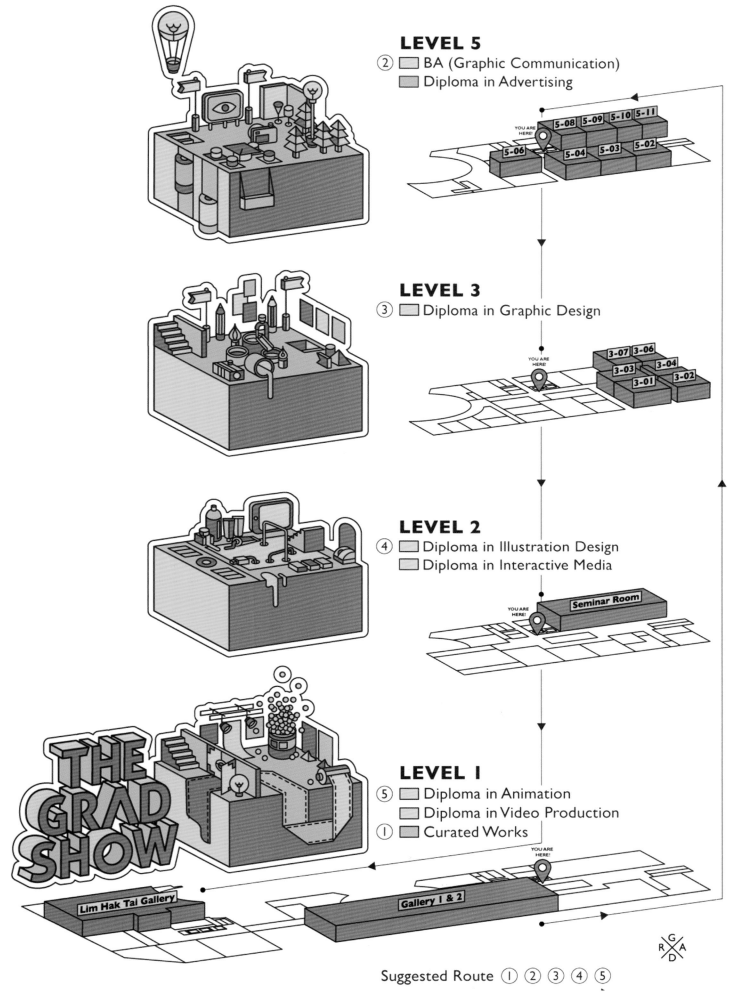

Venice Biennale HK

Design agency: Hybird
Country: China

The Hong Kong Institute of Architects and Hong Kong Arts Development Council participated in the 13th International Architecture Biennale in Venice, curated by oval partnership. Titled "Inter Cities / Intra Cities: Ghostwriting the Future", the theme explores the sustainable urban growth for the future of Hong Kong and large cities around the world in accordance with the biennale's chief curator David Chipperfield's topic of "Common Ground". It will assess the process of sustaining an increasing population in a globalised commercial economy, while fostering a sense of cultural identity. The exhibit will specifically analyse the 320 hectare district known as kowloon east, its officially and unofficially planned development projects and the scope of its existing and future culture, making it one of the most pertinent regeneration projects in the world.

威尼斯双年展香港展区

香港建筑师协会和香港艺术发展局参加了第13届威尼斯国际建筑双年展。以"城市之间：为未来代笔"为主题的展览，探索了香港及全球各大城市未来的可持续城市发展，符合本次双年展主策展人大卫·奇普菲尔德的主题"共同基础"。展览将评估全球化商业经济体是如何面对与日俱增的人口的，从而营造出一种文化认同。

设计师：Hybird 设计公司　国家：中国

Art

Le Corbusier: An Atlas of Modern Landscapes Exhibition

Design agency: Department of Graphic Design and Advertising, MoMA
Creative director: Hsien-yin Ingrid Chou
Designer: Hsien-yin Ingrid Chou, Samuel Sherman, Tony Lee
Typography: Samuel Sherman, Tony Lee, Tsan Yin
Photographer: Martin Seck
Country: USA

Le Corbusier: An Atlas of Modern Landscapes is MoMA's first major exhibition on the work of Le Corbusier encompassing his work as an architect, interior designer, artist, city planner, writer, designer and photographer. The exhibition reveals the ways in which Le Corbusier observed and imagined landscapes throughout his career, using all the artistic techniques at his disposal, from watercolours to sketches, from the photographs to the models, and from illustration to sculpture. It was a very clear to the designers that they should created a custom font inspired by his typography for this exhibition. The result reveals a cohesive visual impact for an artist who embraced cross-disciplinary practice.

勒·柯布西耶：现代景观地图展览

"勒·柯布西耶：现代景观地图"展览是纽约现代美术馆第一次举办有关勒·柯布西耶的大型展览，内容覆盖了他作为建筑师、室内设计师、艺术家、城市规划师、作家、设计师和摄影师的方方面面。展览展示了柯布西耶如何通过自己的事业来观察和想象景观。从水彩到素描，从摄影到模型，从插画到雕塑，他运用各种艺术手段来呈现景观。设计师认为他们应当为展览专门设计一套字体。最终的设计与柯布西耶的跨学科实践十分匹配。

设计机构：MoMA 平面设计与广告部 创意总监：希因·英格丽德·周 设计师：希因·英格丽德·周、塞缪尔·谢尔曼、托尼·李 字体设计：塞缪尔·谢尔曼、托尼·李、珊·殷 摄影：马丁·塞克 国家：美国

CHANDIGARH: A NEW URBAN LANDSCAPE FOR INDIA
1945–65

After World War II, Le Corbusier faced new frustrations, notably when the United Nations Headquarters in New York was seen to completion by Wallace K. Harrison. But a chance to design an entire city finally came, in 1950, with the commission for a new regional capital at Chandigarh, in the northern Indian state of Punjab. This was one of the most monumental realizations of the new poetics of raw concrete and a chance to design over a vast landscape, implementing visual schemes the architect had first encountered in his studies of ancient Rome, three decades earlier. On twice-yearly flights between Europe and India, he partook in "the view of the airplane," looking down on the varied landscape—views he recorded in his sketchbooks.

His sculpture in those years echoed his architecture, from works in wood to sand casts developed on Long Island. He also continued to write, publishing numerous books. He introduced the Modulor, his system of harmonic proportions, in 1947, and on behalf of the "Synthesis of the Arts"—integrating architecture, painting, and sculpture—he strove to become the central figure of a modern architecture by then almost universally accepted.

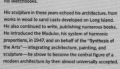

RESPONDING TO LANDSCAPE, FROM AFRICA TO THE AMERICAS
1929–40

The international success of Le Corbusier's books led to invitations to travel and thus to encounters with new landscapes. The architect's first major success outside Switzerland and France came in 1928, with the commission for the headquarters of the Central Union of Consumer Cooperatives, or Centrosoyuz, in Moscow. Dreaming of building on a large scale with an urban impact, Le Corbusier was stung by his defeat in the competition for the Palace of the Soviets in Moscow in 1932, and he transferred his hopes for commissions to Fascist Italy, though without success.

In 1929 he developed plans for Rio de Janeiro, São Paulo, and Montevideo, inspired by impressions gained when flying over those cities, but even the enthusiastic welcome of the local elites did not lead to their implementation. Likewise, he campaigned tirelessly, but in vain, to realize his iconoclastic plans for the transformation of Algiers.

Lectures were a key means by which Le Corbusier persuaded the public of the validity of his approach, a demonstration made palpable by his technique of drawing on long scrolls of paper while lecturing in North and South America, examples of which are presented here.

FROM THE JURA MOUNTAINS TO THE WIDE WORLD
1887–1917

Le Corbusier was born Charles-Édouard Jeanneret on October 6, 1887, in La Chaux-de-Fonds, in French-speaking Switzerland. The city was the world center for watchmaking, and his parents aspired for him to become an engraver of watchcases. He learned to draw, exploring the landscape of the Jura Mountains, before focusing on architecture. At age twenty he built his first house, Villa Fallet, in the hills above the town center.

Over the next five years his horizons expanded to the edges of Europe, always animated by the dialogue between tradition and modernity. In 1907 he took a study trip to Italy before heading to Vienna. He worked in the Paris architectural studio of Auguste Perret, pioneer in the use of reinforced concrete, then set out to study city planning in Germany; in Berlin he worked in the studio of architect Peter Behrens. In 1911 he took his "Journey to the East," through the Balkans and Istanbul to Greece. Back home, he taught architecture and interior design and built several houses, drawing upon the landscapes and modern practices he had observed in Vienna, Paris, and Berlin.

Barcelona Direccions Exhibition

Design agency: clasebcn design
Client: Ajuntament de Barcelona
Country: Spain

Both the exhibition graphics and the communication campaign are based on a graphically simple yet very well-known, dynamic idea: a huge array of arrows all pointing in the same direction to reflect a city that never stops and marches ever on. All the graphics are created with two basic colours to deliver the brightest, most optimistic Mediterranean message.

巴塞罗那导视展

展览的图形设计和宣传活动都以简单却知名的概念为基础：一排巨大的箭头全部指向同一方向，反映了城市永不停息、永远向前的精神。所有图形都采用两种基本色调，传递出明媚、向上的地中海讯息。

设计机构：clasebcn 设计公司 委托方：巴塞罗那市政厅 国家：西班牙

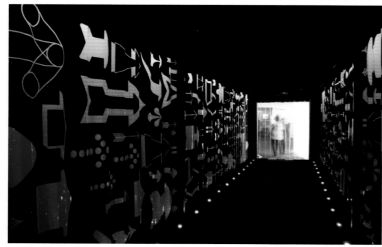

Art

Secrets of the Royal Bedchamber

Design agency: Bibliothèque design
Client: Historic Royal Palaces
Country: UK

An exhibition at Hampton Court Palace, showcasing the largest collection of State beds from the 17th and 18th century royal court. Secrets of the Royal Bedchamber explores why and how the bedchamber became the most important and public destination in the palace. The schemes contemporary interpretation of Baroque design, utilises principles of reflection, embellishment and symmetry. Working with the restrictions of a Grade I listed building, Bibliothèque developed a large freestanding signage system, to compete with the impressive scale of the rooms. Rich colours were used to aid navigation through the monochrome exhibition, while a combination of materials and unusual processes echoed the lavish design of the period.

皇家寝室的秘密

本次展览位于英国汉普敦皇宫内，展示了17世纪和18世纪英国皇室最大规模的寝具收藏。"皇家寝室的秘密"展览探索了寝室是如何成为皇宫中最受欢迎的景点以及它们的重要性。设计方案以现代的方式诠释了巴洛克式设计，充分利用了反光、装饰和对称等设计法则。在一级保护历史建筑的严格限制下，Bibliothèque设计公司开发了一套大型独立式导视系统，与房间的宏大规模十分相配。丰富的色彩有助于在单个展览中辅助导航，而材料的组合和与众不同的流程则与当时奢华的设计相互呼应。

设计机构：Bibliothèque设计公司 委托方：历史皇家宫殿组织 国家：英国

Art

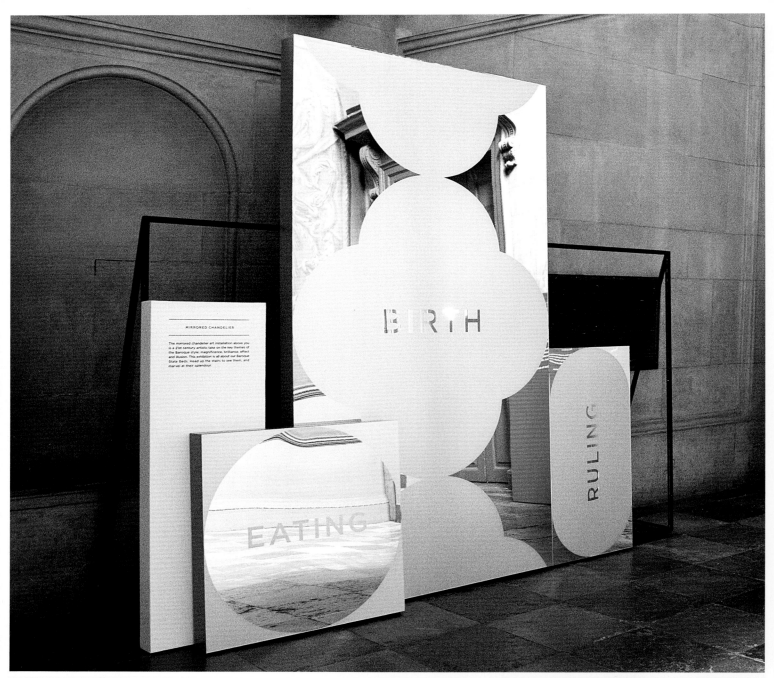
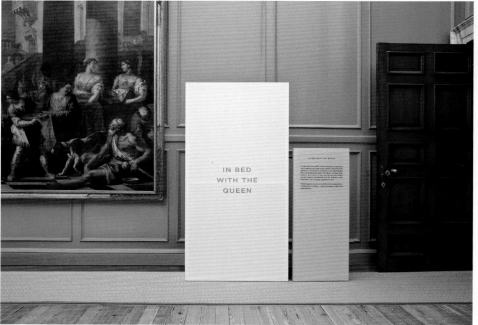

Art

Garden and Spring

Design agency: SALU.io
Designer: Michael Salu
Photographer: Vipul Sangol
Client: The Abraaj Group/Art Dubai
Country: UAE

SALU was commissioned to develop a fully integrated visual identity, environmental and digital experience for The Abraaj Group Art Prize 2014. The design of Garden and Spring takes its inspiration from the foundation of the Indo-Islamic garden. A simple architectural grid provides the basis for every visual element of the project, from the bespoke typeface and signage to the webapp http://gardenandspring.com/.

花园与春天

SALU 受委托为 2014 阿布拉吉集团艺术奖设计全套的视觉识别、环境及数字体验系统。"花园与春天"的设计从印度－伊斯兰花园中获得了灵感。简单的建筑网格为每个视觉元素提供了基础，覆盖了从定制的字体和标识到网站应用（http://gardenandspring.com/）的全部设计。

设计机构：SALU 设计公司 设计师：迈克尔·萨卢
摄影：维普尔·桑古尔 委托方：阿布拉吉集团 /
艺术迪拜 国家：阿联酋

Bagh O Bahar:
GARDEN AND SPRING

THE ABRAAJ GROUP

The Abraaj Group Art Prize 2014

Exhibition Continues ▶▶

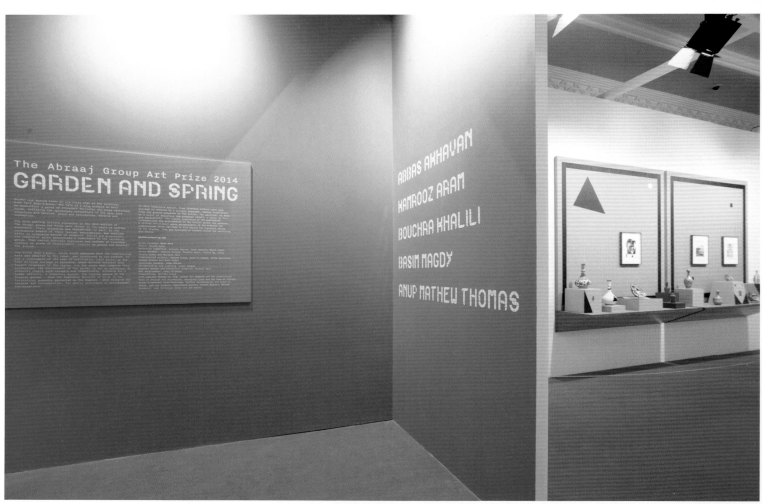

ABBAS AKHAVAN

KAMROOZ ARAM

BOUCHRA KHALILI

BASIM MAGDY

ANUP MATHEW THOMAS

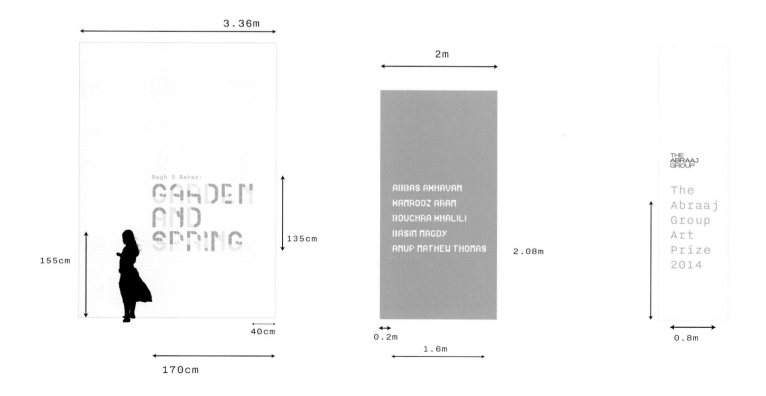

D&AD Awards Ceremony 2013

Design agency: Bravo Charlie Mike Hotel
Photographer: John Hooper
Client: D&AD
Country: UK

Bravo Charlie Mike Hotel were commissioned to design and art direct the event identity for the 2013 Awards Ceremony. The design idea highlighted the superlative nature of the award-winning work and people involved. With the right '&' the designers were able to transform 'D&AD' into a very colloquial English use of the word 'DEAD', meaning 'really'. For example, 'dead excited', 'dead jealous', 'dead chuffed', 'dead nervous', etc. This allowed the designers to illustrate the heightened emotions and excitement felt in the lead up to and during the event but with a bit of humour because, after all, it's a celebration. The flexibility of the concept also meant that it could be used very effectively and responsively in different contexts to unite all of the areas and situations in which the identity would be needed. Dead simple.

2013 D&AD 颁奖礼

Bravo Charlie Mike Hotel设计公司受委托为2013D&AD颁奖礼进行活动设计。设计概念突出了获奖作品和参与者的特色。设计师利用花体的"&"符号，将"D&AD"变成了"DEAD"，在英语口语中有"特别，真的"的意思。这让设计师得以表现颁奖活动中高昂的情绪和兴奋感，同时也兼具一些幽默感。设计概念的灵活性保证了它可以被应用在不同的环境和场合中，十分简洁便利。

设计机构：Bravo Charlie Mike Hotel设计公司 摄影：约翰·胡泊尔 委托方：D&AD 国家：英国

Design Museum – Sustainable Futures* Exhibition

Design agency: Build
Designer: Michael C. Place
Client: The Design Museum
Country: UK

An identity, custom typeface, exhibition graphics and all supporting material for a travelling exhibition about sustainability, produced by Build. Given its subject matter and travel requirements, special consideration also had to be given to the materials and processes used for the production of all elements of the project.

设计博物馆——可持续未来展览

Build设计公司为这次巡回展览提供了形象识别、定制字体、展览图形和所有辅助材料的设计。根据展览主题和巡回展出的要求，设计师特别注重了项目中所有元素的制作材料和制作流程的选择。

设计机构：Build设计公司 设计师：迈克尔·C·普雷斯 委托方：设计博物馆 国家：英国

CITIES

—Cities are often regarded as hugely damaging resource hungry polluters. Many argue, however that cities are more sustainable than rural areas because of their density. Urban dwellers live, work and go to school within close proximity to one another, with easy access to local amenities. Their footprint on the land is smaller, in some cases people live in high-rise housing and community living. Planners call it 'access by proximity' and is deemed a more sustainable way of living than people residing in single houses, and relying on a car to do everyday activities.

—There are great examples of existing cities or developing settlements that take up the challenge of working towards a more sustainable infrastructure, where planners, architects, designers and the citizens take their role seriously in terms of challenging traditions and behaviours.
From architects responding to natural disasters, to examples of independent buildings and communities; these projects provide a unique insight into how we can use design and architecture to communicate viable principles of construction, engineering, use of materials and natural resources, to promote real solutions.

"SOLAR POWER IS NOT ABOUT FASHION, IT'S ABOUT SURVIVAL"
LORD NORMAN FOSTER

Oriental Watertown

Design agency: Hstudio
Designer: Alessandro Antonuccio
Client: San Servolo Servizi
Country: Italy

"Oriental Watertown, Photos from the City of Suzhou" is an exhibition of Chinese photographers who have portrayed the city of Suzhou, twin city of Venice. Hstudio designed the graphic image of the event and applied it on various media, including the portal that welcomes visitors and on which all the artists and organisers were listed.

东方水城

"东方水城——苏州城市摄影展"展出了威尼斯的姐妹城市——苏州的城市摄影。Hstudio 为该活动提供了图形设计并将其应用到了各种媒介上，包括列出了所有艺术家和组织者名单的迎宾拱门。

设计机构：Hstudio设计公司 设计师：亚历山大·安东努西奥 委托方：圣塞尔沃洛服务公司 国家：意大利

SO 100 Exhibition / SO Architecture

Design agency: SO Architecture / Shachar Lulav & Oded Rozenkier
Photographer: Shai Epstein
Country: Israel

The exhibition deployed and analysed the office view about 13 different concepts. The location of the different concepts was set by a diagram that determined the contexts between each topic to the other topics. The designers defined three basic concepts that they see as essential in all of their architectural work: contemporary, ideology and aesthetics. The three spaces where the concepts were being examined were wider in accordance with the importance that the designers give them. Each space presented an essence of a theoretical text that described the designers' reference toward the discussed concept. In addition to the text, 3 panels per concept were presented. In these panels the discussed concept was being reflected through selected projects. In addition to these three basic concepts, the designers defined 10 more topics that generate their design thinking in all of their work. These width topics were using as parameters for creating and analysing their work. In most of the projects, the same topics are being implemented in a different mix from project to project. The exhibition design created an experience where the designers directed the visitor's viewing perception using free movement on mobile office chairs. The various exhibits occupied a three-dimensional grid above the viewer's head, and in an analogy to urban flaneury, the visitor experienced the exhibition in an unusual angle while moving about freely. Similarly to an urban excursion, the visitor may enter more intimate spaces where theme presentations of the works were presented. Such themes included the handling of light, space, scenario, materiality, section, and other issues underlying their architectural work.

SO 建筑事务所 100 展览

展览展开并分析了 SO 建筑事务所的 13 个设计主题。不同主题的位置由一个图表决定,图表明确了不同主题之间的关系。设计师定义了三个基本建筑主题:现代、意识形态和美学。这三个主题由于其自身的重要性所占的空间更大。每个空间呈现了一个理论文本的概要,描述了设计师对该主题的参考说明。除了文本信息,每个主题还设有三块展板。这些展板通过精选的项目对该主题进行了探讨。除了三个基本主题外,设计师还定义了 10 多个与设计相关的主题。设计师通过这些主题对它们的作品进行了探讨和分析。大多数项目都混合了多个主题。展览设计让设计师可以通过办公转椅的自由移动来引导参观者的视线。各种各样的展览品占用了参观者头顶的立体网格空间。参观者以独特的视角体验展览,同时又能活动自如。与城市短途旅行相似,参观者可以深入作品主题展示的更深处进行参观。这些主题包括光的处理、空间、场景、材料、剖面以及其他与建筑相关的元素。

设计机构:SO 建筑事务所 /SL&OR 设计公司 摄影:沙伊·爱泼斯坦 国家:以色列

Art

Varini Nell'autosilo

Design agency: CCRZ
Designer: Marco Cassino, Paolo Cavalli
Photographer: CCRZ
Client: LAC Lugano Arte Cultura / Città di Lugano
Country: Switzerland

Swiss artist Felice Varini creates three permanent works in the LAC underground parking, the first being realised for the new art museum's collection. The exhibition is open to pedestrians for five days before the car park's inauguration. The communication elements include website, poster, invitation and indoor signage. An orange cube is installed over the entrance of the car park to signal the exhibition. A TV commercial is produced from Patrik Soergel's documentary on the realisation of Varini's works.

瓦里尼的车库展览

瑞士艺术家菲利斯·瓦里尼在卢加诺艺术文化馆的地下车库内打造了三件永久性作品，这是这座新艺术博物馆的第一批藏品。展览在停车场正式使用前面向行人开放五天。展览的传播元素包括网站、海报、邀请函和室内标识。车库门前安装了一个橙色立方体作为展览的标志。帕特里克·泽格尔拍摄了瓦里尼作品创作的纪录片，并且在电视上播放了广告。

设计机构：CCRZ 设计师：马尔科·卡西诺、保罗·卡瓦利 摄影：CCRZ 委托方：卢加诺艺术文化馆/卢加诺市 国家：瑞士

Stripes

Design agency: Stockholm Design Lab
Designer: Lisa Fleck
Client: Nordiska museet
Country: Switzerland

Solution
Undaunted by the Nordiska Museet's august, highly ornate architecture, the designers created an identity that was unashamedly contemporary and bold and provided a dramatic visual contrast, with bright, broad diagonal bands of colour and clear, modern, sans serif typography. From posters and ads to the enormous hoardings at the front of the museum to the exhibition set in the central hall, the effect was startling. For the catalogue, rich with imagery, abstract pattern and essays by leading writers, Stockholm Design Lab designed unique covers for all 1500 printed copies.

Result
The exhibition was greeted with overwhelmingly positive reviews and brought a different kind of visitor to the Nordiska museet, one not so attracted by its usual output of dry, rather academic exhibitions. The designers also picked up a prestigious Guldägget (Golden Egg) for the identity in the 2014 awards.

条纹

方案
在北欧博物馆华丽的古典建筑背景下，设计师打造了大胆、现代的标识系统，形成了巨大的视觉反差。设计选择了亮色斜条纹和简洁现代的无衬线字体。从海报、广告页到博物馆门前巨大的广告牌，整个设计的效果十分出众。展览名录中收录了丰富的抽象图形和领衔作者撰写文章，Stockholm Design Lab 为 1500 本名录设计了独特的封面。

成果
展览获得了广泛的好评，为北欧博物馆带来了不同类型的参观者。他们原本对博物馆平常的学术性展览并没有什么兴趣。设计师由于这次展览设计在 2014 年获得了大量奖项。

设计机构：Stockholm Design Lab 设计公司 设计师：丽萨·弗莱克 委托方：北欧博物馆 国家：瑞士

MCP / Museo Civico Prato

Design agency: D'Apostrophe with Lcd
Creative director: Donatello D'Angelo / Gianni Sinni
Designer: Cosimo Damiano D'Aprile / Laurie Elie
Client: Comune di Prato (Italy)
Country: Italy

The font design is a reinterpretation of contemporary rotunda Gothic script in use in Italy at the time of Francis Datini; from this reworking has resulted logo, formed by the Museum of Prato, MCP, synthesised in the use of only the vertical legs of the letters; Lcd designed the institutional font "Datini MCP", which together with the extensive family of Fedra Serif A, Sans and Mono, will contribute to the definition of all typographic communication of the Museum; the colour palette uses three soft colours (plus black and white) – gold, aviation, wine – that do not compete with the images of the works; the dynamism of the project is ensured by the use of elements in constant evolution / mutation, such as colour and pattern system adaptability content and localisation of the spaces of the exhibition.

普拉托市图书馆

字体设计是对现代圆体 Gothic 字体的重新诠释，由此衍生了普拉托市图书馆的标志，该标志仅采用了字体细长的"腿"。Lcd 设计了规范字体 Datini MCP，该字体与 Fedra Serif A、Sans and Mono 共同组成了博物馆的所有文字传播设计。设计的色彩搭配选用了三种柔和的颜色（除黑白两色外）——金色、天蓝色、酒红色，它们不会影响艺术品的展出效果。不断演变的元素保证了项目的动态性，其中包括图案系统的适应性内容、展览空间的本地化等。

设计机构：D'Apostrophe 设计公司、Lcd 设计公司 创意总监：多纳泰罗·德安吉洛、詹尼·辛尼 设计师：科西莫·达米亚诺·达普利勒、劳里·艾利 国家：意大利

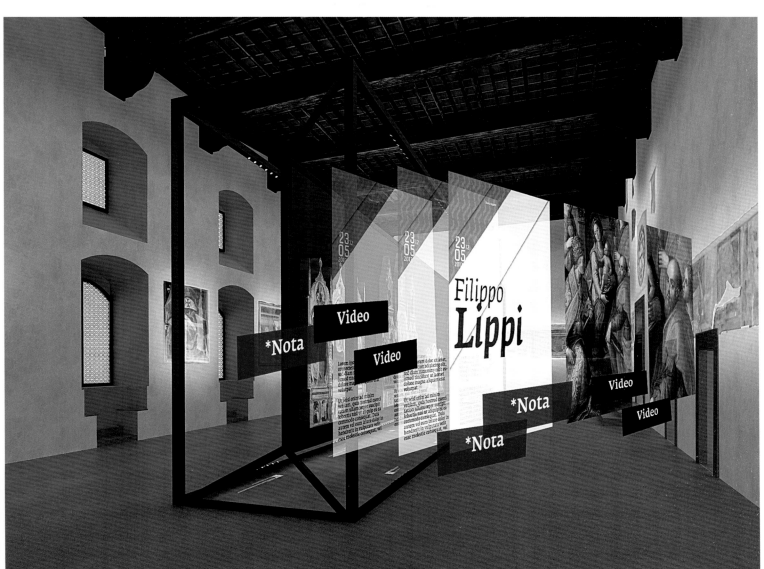

AD13 Navigate

Designer: Emme Jacob
Photographer: Sean Wakely
Client: AUT University
Country: New Zealand

AUT University held its annual Art + Design Festival in a new space in 2013. The space provided a dynamic and artistic backdrop to the show, but also presented new issues in finding your way throughout the event. A way-finding system was designed to lead guests between each show and through each individual show with ease.

AD13 导航

2013年，奥克兰理工大学举办了一年一度的艺术设计节。展览空间提供了一个充满活力和艺术气息的背景，但是对整个活动的导航提出了新的挑战。该导视系统能帮助宾客在各个展览内部和之间穿梭自如。

设计师：埃姆·雅各布 摄影：肖恩·维克利 委托方：奥克兰理工大学 国家：新西兰

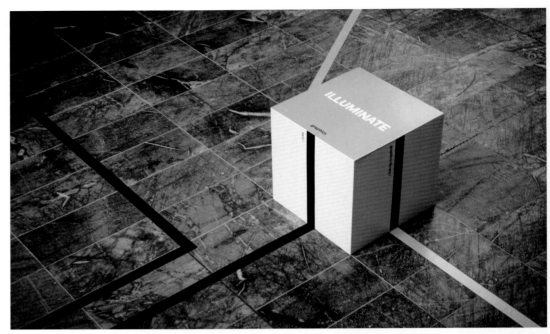

The Art of Scent 1889-2014 in Madrid

Design agency: Cano Estudio
Country: Spain

The Art of Scent 1889-2014 in Madrid is the first exhibition that shows and analyses scent. The exhibition is curated by Chander Burr, former perfume critic of New York Times. Cano Estudio is commissioned with the catalogue and exhibition design. "We are looking for a space where visitors will hold their breaths and forget about the outside world for a moment. The goal is to create a relaxing atmosphere to feel the scent, a pure and natural scent without any modification or container. Simply, there is only scent, thinking and feeling. This is to create a space which is visually empty yet attractive," explained Jesús Cano, the exhibition designer. "We create a blank canvas and paint on the board of sense with light. In this atmosphere, we can focus on our most neglected sense, sense of smell. Because we want the sense of smell to be next to our emotions," Jesús Cano added. Cano Estudio also designed the exhibition's graphic design. In addition, this exhibition explores the relation between each scent and iconic work in art history. Visitors can discover representative work of each era through QR code, just as these scents.

气味的艺术 1889-2014 马德里

气味的艺术 1889-2014 是首个把气味作为艺术品来展示和分析的展览。此展览是由前《纽约时报》的香水评论家钱德勒·波尔负责策划，并由 Cano Estudio 负责编目和陈列的。"我们寻求一种能够内在地让参观者屏息静气，片刻地忘掉外部世界的空间。目标是创造出一种放松的氛围来感受香气。一种最纯粹状态的，没有任何修饰，也没有任何容器和包装的香气。简单的只有香气、思想和感受。这是要创造一个即便不在视觉上做任何事并且实质上全空却也能够很吸引人的空间。"负责陈列设计的朱瑟斯·卡诺这样说。"我们创造出一幅空白的画布，用光在感觉的面板上作画，来作出能让我们集中感受其中一个最被忽略的感官——嗅觉——的这样一个气泡。因为我们想要提醒嗅觉实际上是离情绪最近的。"朱瑟斯·卡诺这样补充道。Cano Estudio 还负责了气味的艺术 1889-2014 展览的图形图案设计。另外，本次展览还揭示了每种香气与艺术史上的标志性作品之间的关系。通过一个二维码，就可以发现每个时期的代表性艺术作品，就像这些香气一样。

设计机构：Cano Estudio 设计公司 国家：西班牙

SURREALISMO

Angel
1992

OLIVIER CRESP
CEDIDO POR THIERRY MUGLER, CLARINS

El autor Thomas Pynchon escribió que el Surrealismo combinaba «dentro de un mismo marco, elementos que normalmente no están juntos para provocar efectos sorprendentes y carentes de lógica». Olivier Cresp ilustró este principio con Angel, su obra olfativa surrealista más influyente. Diseñó una estructura en trípode, combinando tres materiales que no se suelen encontrar juntos: la cumarina, una molécula similar al mazapán; una hierba terrosa, húmeda y tropical; y etil maltol, la molécula que confiere el olor dulzón al algodón de azúcar. El resultado fue una obra sorprendente y carente de lógica, con una estética que la hace realmente extraordinaria. Parecía como si Cresp lo hubiera vuelto todo mucho más auténtico. Su uso del etil maltol como materia estructural en lugar de sutil ornamento redefinió las fronteras del arte olfativo e hizo que la asombrosa artificialidad quedara en primer plano.

SURREALISM

Angel
1992

OLIVIER CRESP
BY THIERRY MUGLER, CLARINS

Pynchon wrote in 1984 that Surrealism combined "in the same frame elements not normally found together to produce illogical and startling effects". Artist Olivier Cresp illustrated this with Angel, his 1992 work of olfactory Surrealism. Cresp designed a tripod structure, combining three materials not usually found together: the marzipan-like molecule coumarin; an earthy, humid tropical grass; and ethyl maltol, the synthetic molecule that gives cotton candy its synthetic sweet smell. The result of these elements was an illogical, startling work, whose aesthetics made it extraordinary. It seemed as if Cresp suddenly made everything more authentic. His use of ethyl maltol, which he transformed from subtle ornament to fundamental structural material, took olfactory art to new extremes and placed its artificiality in full view.

For A World of Its Own

Design agency: Department of Graphic Design and Advertising, MoMA
Art director: Greg Hathaway
Creative director: Mike Abbink, Hsien-yin Ingrid Chou
Designer: Luke Williams
Photographer: Martin Seck
Country: USA

A World of its Own: Photographic Practices in the Studio is an exhibition at MoMA that highlights photographic artworks from the last century, observing the multitude of ways that photographers have used the space, lighting, and environment of their studios to achieve a desired image. Inspired by the exhibition's theme of controlling an environment, the designers designed a title treatment that activates and challenges the exterior walls of the galleries, inviting the visitor to perceive an imaginary space; in essence, positioning them inside the photographer's studio.

自己的世界

"自己的世界：工作室摄影展"是纽约现代美术馆的一次展览。展览重点展示了20世纪的摄影艺术品，分析了摄影师是如何利用他们工作室的空间、灯光和环境来创作出优秀的图像作品的。设计师从展览的主题——控制环境中获得了灵感，设计了一个能够活跃并挑战展厅外墙的标题，邀请参观者来体验图像空间。从本质上来说，展览将参观者引到了摄影师的工作室内部。

设计机构：MoMA平面设计与广告部 艺术总监：格雷格·哈塞维 创意总监：迈克·阿宾克、希因·英格丽德·周 设计师：卢克·威廉姆斯 摄影：马丁·塞克 国家：美国

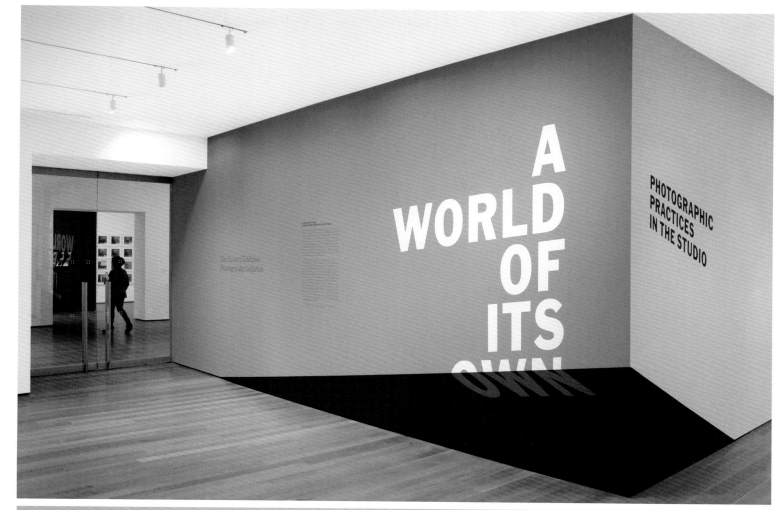

Gauguin: Metamorphoses

Design agency: Department of Graphic Design and Advertising, MoMA
Creative director: Mike Abbink, Hsien-yin Ingrid Chou
Designer: Tony Lee
Photographer: Tony Lee, Martin Seck
Country: USA

This exhibition focuses on these less well-known aspects of Gauguin's rare and extraordinary prints he created in several discrete bursts of activity from 1889 until his death, in 1903. These remarkable works reflect his experiments with a range of techniques, from bold, rough-hewn woodcuts to jewel-like watercolour monotypes to evocative transfer drawings. In comparison with his paintings, Gauguin's prints tend to be darker, more indefinite, or more abstracted. He was drawn to printing techniques that engendered subtle or blurred textures, nuanced colours, and accidental markings, all of which impart a mysterious, dreamlike quality to his images. The designers designed the title wall to reflect his exceptionally experimental and mysterious prints.

高更：变形记

展览聚焦于高更鲜为人知的一面，展示了他从1889年至1903年所创作的一系列珍稀的印制作品。这些非凡的作品反映了他对技术的广泛试验，从粗犷的木刻、贵重的水彩单型刻印到能够引起共鸣的转移画。与画作相比，高更的印制作品更昏暗、模糊、抽象。他深受印制技术微妙模糊的纹理、细致的色彩和偶然的记号所吸引，这些元素给予了他的图像神秘、梦幻的感觉。设计师在主题墙的设计中反映了高更非凡而神秘的印制作品的特色。

设计机构：MoMA 平面设计与广告部　创意总监：迈克·阿宾克、希因·英格丽德·周　设计师：托尼·李　摄影：托尼·李、马丁·塞克　国家：美国

Art

Material Art Fair

Designer: SAVVY STUDIO
Client: Mexico City
Country: Mexico

The premise behind the expositive design for Mexico City's Material Art Fair was to use the same material resources and reinterpret them in each one of its applications, starting from a module that grows and transforms at the same time as elements are added up. This way, using simple materials like planks or slim strips of natural pine, a series of modules with different functions were created. Inspired in the modular design of Sol Lewitt that allows the reproduction of countless pieces from a simple structure, the main module is made out of strips of wood that form geometrical figures such as triangles and squares. This structure holds the fair's official poster and is thought out to change shape and evolve along with the event, adding up more modules edition after edition. To divide some spaces, following the same concept of fine lines and modular evolution, overlapped planks are placed in different angles to create empty spaces with symmetrical shapes, thus simulating the patron of cobblestone walls.

材料艺术展

墨西哥城材料艺术展的说明设计选用了与展览材料相同的材料,并且对它们进行了重新诠释。设计从一个模块开始,随着元素的累加而不断成长和变化。这样一来,使用木板细松木条等简单的材料就能打造一系列模块。设计师从勒维特可不断复制的模块设计中获得了灵感,主模块由组成三角形和正方形的木条构成。这个结构上悬挂了展会的官方海报,并可以随着活动的开展而变换形状和进化。为了分割空间,设计师将模板以不同的角度叠加起来,营造出对称形状的空白空间,与圆石墙形成了相互映衬。

设计师:SAVVY 工作室 委托方:墨西哥城 国家:墨西哥

Soundings: A Contemporary Score

Design agency: Department of Graphic Design and Advertising, MoMA
Creative director: Hsien-yin Ingrid Chou
Art director: Greg Hathaway
Designer: Luke Williams
Production: Paulette Giguere, Tom Black
Photographer: Martin Seck, Luke Williams
Country: USA

Sounding: A Contemporary Score is an exhibition about innovative contemporary sound art. These artistic responses range from architectural interventions, to visualisations of otherwise inaudible sound, to an exploration of how sound ricochets within a gallery, to a range of field recordings. Inspired by the artworks, the designers decided to focus on two key aspects of the theme – architecture interventions and sound ricochets. Through large scale of the typography, and the placement of the type, the designers created a visualisation of sound for the exhibition title wall.

回响：现代配乐

"回声：现代配乐"展览展示了创新的现代声音艺术。这些艺术展示包括建筑处理、声效视觉化以及如何让声音在展厅内跳跃和大量实地录音。设计师从艺术品中获得了灵感，决定以设计的两个方面——建筑处理和声音跳跃。通过大尺寸的字体和文字排版，设计师在展览的主题墙上实现了声音的视觉化。

设计机构：MoMA平面设计与广告部 创意总监：希因·英格丽德·周 艺术总监：格雷格·哈塞威 设计师：卢克·威廉姆斯 制作：波莱特·盖古勒、汤姆·布莱克 摄影：马丁·塞克、卢克·威廉姆斯 国家：美国

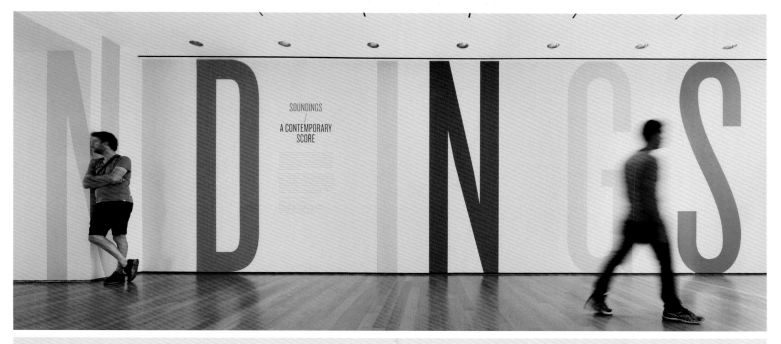

SOUNDINGS / A CONTEMPORARY SCORE

INDEX ≫

索引

Aron Filkey, Nora Demeczky
136

Bailey Lauerman
052

Bibliothèque design
032, 036

Bibliothèque design
198

Bravo Charlie Mike Hotel
066, 072, 206

Build
208

Cano Estudio
226

CCRZ + Decoma Design
020, 164, 218

CCRZ + Studio Brambilla Orsoni
068, 132

Christian Pannicke
158

clasebcn design
196

D'Apostrophe
118, 178, 180, 222

Department of Graphic Design and Advertising, MoMA
184, 194, 230, 232, 236

EDG Experience Design Group
012

Emme Jacob
224

Estudi Conrad Torras & La Petita Dimensió
166

Frits van Dongen - CIE with D'Apostrophe
050

futago_
122, 126, 186

Gresham, Smith and Partners
170

Hstudio
212

Hybird
192

I-AM London
060, 062

ico Design
006

Idéskilte Horsens ApS
174

Ippolito Fleitz Group GmbH
076, 082, 086, 092, 096, 102, 106

karlssonwilker inc.
156

KUKI Krzysztof Iwanski
140

Kxdesigners
016, 046

Lantavos Projects Architecture and Design
112

Lee H. Skolnick Architecture + Design Partnership
034, 150

Liminal Graphics
110

Margot Lombaert
144

nicklashaslestad
138

NOSIGNER
152

Petit Comitè
154

Practice and Workroom
160

SALU.io
202

Sasaki Associates, Inc.
044, 172

SAVVY STUDIO
234

SO Architecture / Shachar Lulav & Oded Rozenkier
214

Stockholm Design Lab
074, 220

Student Committee of The Grad Show
190

THERE
054, 056

ujidesign
038, 040

UMA/design farm
028

UNION ATELIER
142

VID Lab
022

图书在版编目（CIP）数据

展览导视. 2 / （日）前田丰编 ; 常文心译.
沈阳 : 辽宁科学技术出版社, 2015.5
ISBN 978-7-5381-9194-3

Ⅰ. ①展… Ⅱ. ①前… ②常… Ⅲ. ①展览馆－导视设计－世界－现代－图集 Ⅳ. ①TU242.5-64

中国版本图书馆CIP数据核字(2015)第071409号

出版发行：辽宁科学技术出版社
　　（地址：沈阳市和平区十一纬路29号　邮编：110003）
印　刷　者：利丰雅高印刷（深圳）有限公司
经　销　者：各地新华书店
幅面尺寸：215mm×285mm
印　　张：15
插　　页：4
字　　数：30千字
印　　数：1～1500
出版时间：2015年 6 月第 1 版
印刷时间：2015年 6 月第 1 次印刷
责任编辑：周　洁
封面设计：关木子
版式设计：关木子
责任校对：周　文
书　　号：ISBN 978-7-5381-9194-3
定　　价：258.00元

联系电话：024-23284360
邮购热线：024-23284502
http://www.lnkj.com.cn